ECOLOGY

One touch of nature
makes the whole world kin.

—William Shakespeare

Written by
Pamela Spurling Jennett

Illustrated by Diane Valko

A special thanks to Raphael R. Payne, Ph.D. and Professor Paul Kuld,
Department of Biological Sciences, Biola University

Table of Contents

Introduction to Ecology 3
Setting the Stage 4
The Scientific Method 6
The Scientific Processes 7

Unit 1: Foundations of Ecology 8
Key Concepts *overhead reproducible* 9
Community 10
Food Webs 13
Adaptations 15
Extension Activities 17

Unit 2: Water Ecology 18
Key Concepts *overhead reproducible* 19
The Water Cycle 20
Water Pollution 24
Water Conservation 29
Extension Activities 31

Unit 3: Air Ecology 32
Key Concepts *overhead reproducible* 33
The Oxygen/Carbon Cycle 34
Air Pollution and Depletion 37
Air Quality Control 41
Extension Activities 43

Unit 4: Soil and Land Ecology 45
Key Concepts *overhead reproducible* 46
The Nitrogen Cycle 47
Soil and Land Exploitation 51
Soil and Land Conservation 54
Extension Activities 57

Culminating Activities 58
Career Corner 59
Assessment 60
Bibliography 62
Ecology Resources 63
Glossary 64

Introduction to Ecology

Use this book as a guide to explore the world of ecology. Each unit opener includes illustrated key concepts followed by activities and experiments for meaningful, hands-on learning. Students practice the scientific method and scientific processes as they work through the following sections:

* *

Foundations of Ecology

Students examine the life of plants and animals in environmental communities. They explore how living and nonliving things are interdependent and how living organisms must adapt to survive. As students study the major biomes of the earth, they investigate the relationships within a biosphere, an ecosystem, a population, and an organism. Through hands-on activities, students explore the purpose of various animal adaptations.

Water Ecology

Students examine water as an essential element for survival of life on our planet. They investigate the water cycle and experiment with turning salt water into fresh water. They make water filters and try to clean up an oil spill. Based on these experiments and demonstrations, students consider ways to help conserve and protect water sources in their communities.

Air Ecology

Students investigate the path of carbon and oxygen through photosynthesis and respiration. They identify ways pollutants change the quality of our air and explore ways to restore and maintain air quality. Students experiment with changes in oxygen and carbon dioxide levels. They observe and analyze the effects of acid rain.

Soil and Land Ecology

Students trace nitrogen as it moves through its cycle—from the atmosphere, into the soil, and along the food chain. They classify soil types and determine which are best for growing crops. Students perform experiments that display erosion and other actions that affect nutrients in the soil. Through activities illustrating the purposes of land use, students discover their responsibility for using the land wisely.

Setting the Stage

Children have a natural curiosity for their environment. This curiosity can be nurtured through the study of ecology. Knowledge of ecology and environmental protection helps students understand that life is dependent on the well-being of the environment. The following ideas set the stage for a classroom investigation into ecology. These activities provide the starting point from which students will research, organize, and evaluate ecology information.

* *

Science Journals

Staple lined paper into a folded construction paper cover. On the first page, have students create a knowledge chart by writing anything they know about ecology. On the next page, have them record any thoughts, ideas, or questions they have about ecology. They can record their answers, results from experiments, ideas for further study, and responses to journal questions on subsequent pages. Provide frequent opportunities for students to share journal entries.

Learning Center

Designate an area in the classroom as your Ecology Corner where students can work independently during their free time. Include reference materials which have photographs and illustrations pertaining to ecology. Display ongoing experiments in this area so students can monitor their progress. Designate wall space where students can post magazine and newspaper articles related to ecology or the environment. Encourage students to bring in small pictures of animals to mount on squares of construction paper to make card sets. Students can use these cards to launch discussions about the role animals play in their habitat.

What We Know About Ecology	Questions We Have About Ecology
Air pollution is a problem.	What is the greenhouse effect?
The rain forests are in danger.	What can I do to save the earth?
The earth has many ecosystems.	What causes water pollution?

Setting the Stage

Famous Faces in Ecology

There are many famous historical figures involved in ecology, among them—Jacques Cousteau, Henry David Thoreau, John James Audubon, Beatrix Potter, John Muir, and Theodore Roosevelt. Invite students to study a person significant to the history of ecology. They can write biographical reports accompanied by illustrations. After completion of the reports, have students design a time line based on the birth dates of the historical figures. Students can then present their research orally in the time line sequence.

Ecology Library

Use the bibliography of fiction and nonfiction resources on pages 62–63 to begin your ecology library. Add student-authored books to the library as you progress through each unit.

The Scientific Method

Throughout each unit there are hands-on experiments which involve students in the steps of the scientific method. *Science Challenge* questions give students a chance to plan additional experiments. As students raise questions, they can use the Scientific Method reproducible on page 6 to assist them in designing their own experiments.

The Scientific Processes

The eight scientific processes are integrated throughout the activities and experiments in each unit. Students can use the reproducible on page 7 to record and reflect on how they used the processes to reach their scientific conclusions.

Name:

The Scientific Method

Question: Based on your observations, identify the problem you wish to explore. Then ask a clear, specific, testable question.

Hypothesis: Make a prediction or your best guess, as to the solution or answer to your question.

Procedure: Plan the materials you will need and the steps you will take to test your hypothesis.

Results and Conclusions: Record the results of your experiment.

Name:

The Scientific Processes

Observing: What was the most important thing you observed during the science experiment?

Communicating: How did you communicate what you learned?

Comparing: How were the results of your experiment the same or different from the results of your classmates?

Ordering: Did you notice any patterns in your data? What in your experiment reflected order in nature?

Categorizing: What was the most important thing you observed during the science experiment?

Relating: What did you learn from this experiment that was related to a fact you previously knew?

Inferring: Based upon what you learned in your experiment, are there other conclusions you might infer that are not direct observations?

Applying: How might you use what you learned to invent something practical to help our world? To explain how something else works?

Foundations of Ecology

The Web of Life

The web of life connects grass to fish,
trees to turtles, sky to caterpillars.
The web of life connects me to everything that came before,
even to the moment when Earth herself was born.
To break the web of life is to fill the universe with pain.

—*Nancy Wood, 1993*

Key Concepts for Foundations of Ecology

Community—

all plants and animals coexisting in a specific environment

Food Webs—

a system of interconnected food chains that demonstrates energy flow for all organisms in an ecosystem

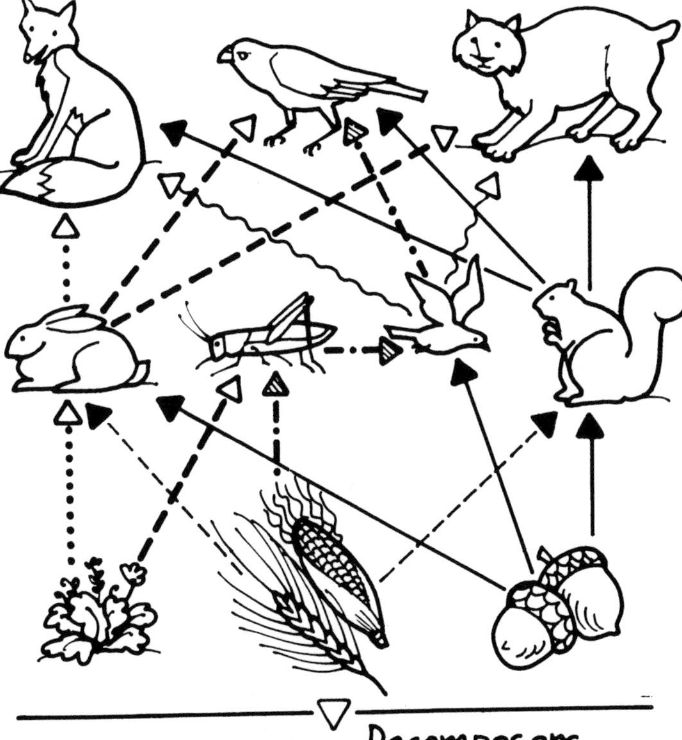

Adaptations—

changes in organisms that improve their ability to survive in their environment

Overhead reproducible

Community

Ecology is the study of the relationships of living organisms to each other and to their environment. In ecology, organisms can be classified into categories such as population, community, biome, and biosphere. **Population** refers to a group of the same organisms living in a common environment. Several populations coexisting in the same environment are referred to as a **community**. Communities coexisting in a major ecological region, such as a desert or tropical forest, form an **ecosystem** or **biome**. All biomes together make up the **biosphere**, the part of planet Earth which sustains life.

In Search of a Community

(Exploration/Recording)

Communities of organisms are all around us. Assign students to observe organisms in an area such as in a rotting log, on grass, or under a rock. Students can record in their journals, making notes about the living and nonliving components in this community.

Types of Biomes

(Researching/Illustrating)

The biosphere, the life-sustaining part of the earth, is divided into six major biomes. These biomes are the arctic tundra, coniferous forest, deciduous forest, grassland, desert, and tropical forest. The climate of an area determines its biome. In cooperative groups, have students study a biome. Ask students to design a zoo exhibit to house animals and plants from that biome.

Range of Tolerance

(Science Experiment)

A range of tolerance is determined by the minimum and maximum requirements that an organism needs to live. Ranges of tolerance can reflect requirements such as the amounts of water an organism needs or the temperatures at which an organism can sustain life. Many organisms remain dormant until their range of tolerance is met, at which time they grow and thrive. In Experiment #1 on pages 11 and 12, students can determine the brine shrimp's range of tolerance for salt in water by hatching and growing shrimp in a variety of saltwater concentrations. The greatest number of eggs will hatch in lower salt concentrations while higher salt concentrations will yield few or no shrimp.

Life Cycle of Brine Shrimp		
Dry Egg	Egg in Salt Water	Hatching Out
Larva	Adult Male	Adult Female

Name:

Range of Tolerance
Experiment 1

EXPERIMENT **1**

Question: *What range of tolerance do brine shrimp have for salt in water?*

Hypothesis: (Choose a range using 0, 2, 4, 6, or 8 teaspoons of salt per one cup of water.)

Materials

- ❑ brine shrimp eggs
- ❑ five clear plastic cups (10 oz. or larger)
- ❑ plastic wrap
- ❑ rubber bands
- ❑ water
- ❑ kosher salt
- ❑ measuring cup
- ❑ five cups of tap water that has been left out overnight
- ❑ craft stick marked ½ inch from one end
- ❑ craft sticks

Procedure:

Step 1

Label the cups *no salt*, *2 tsp. salt*, *4 tsp. salt*, *6 tsp. salt*, and *8 tsp. salt*. Fill each cup with ¾ cup of water.

Step 2

Add the amount of salt stated on the label to each cup of water.

Step 3

Fill the craft stick with eggs to the line on the stick. Gently swirl the eggs into a cup. Repeat this procedure for each cup.

Step 4

Cover each cup with plastic wrap secured with a rubber band. Place the cups in a spot where they will be undisturbed for three days.

Step 5

Observe the cups for hatched eggs. The shrimp are very small, feathery-looking creatures that will be moving.

Note: The shrimp in the illustration are larger than actual brine shrimp.

***Note:** *Brine shrimp eggs are available from a fish supply or biological supply company.*

Name:

Range of Tolerance
Experiment 1

EXPERIMENT 1

Results and Conclusions:

1. Compare the cups. What observations do you make? ____________

2. How do your results support your hypothesis? ____________

3. What are the similarities and differences of your cups compared to the cups of a classmate? ____________

4. Based on this experiment, what does the brine shrimp's range of tolerance for salt in water seem to be? ____________

5. What factors in nature would change the levels of salt in the water so that brine shrimp could or could not hatch? ____________

Science Challenge: Set up an experiment to test this question:

Can I change the salt water in the unhatched cup(s) so that the brine shrimp will hatch?

Write your question, hypothesis, procedure, and materials list on another sheet of paper. Then test the hypothesis and record your conclusions.

Food Webs

Food gives life-sustaining energy to an organism. Organisms obtain energy from food by eating other organisms. The order in which food energy is transferred through several organisms in an environment is called a **food chain**. Organisms in a food chain can be classified into three general categories—**producers, consumers,** and **decomposers**. Most ecosystems have many producers, consumers, and decomposers that are part of several food chains. A system of interconnected food chains that work to provide energy for all organisms in an ecosystem is called a **food web**.

* *

A Link in the Food Chain

(Researching/Classifying)

All living organisms in a food chain can be classified as a producer, consumer, or decomposer. Producers consist mainly of plants. Most consumers are herbivorous, omnivorous, and carnivorous animals. Decomposers, mainly bacteria and fungi, break down dead plants and animals to make nutrients in the soil. Have students refer back to their research on a particular biome in order to make a chart similar to the one below. Invite them to identify examples of producers, consumers, and decomposers native to the biome they researched.

Grass	**Producer:** makes its own food
Mouse	**Herbivore:** a consumer that eats plants
Snake	**Omnivore:** a consumer that eats plants and animals
Hawk	**Carnivore:** a consumer that eats other animals
Bacteria or Fungi	**Decomposer:** an organism that obtains its energy by breaking down dead or decaying material

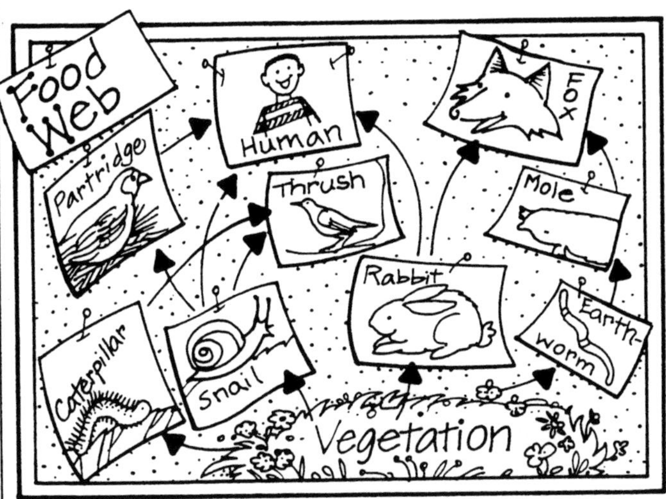

Food Webs

(Critical Thinking)

Explain to students that a series of interconnecting food chains make up a food web. Students can diagram a food web based on their research of biomes. Encourage students to label producers, consumers, and decomposers in their diagrams. Use arrows to show how energy is transferred among organisms in the food web. (See illustration.)

Food Webs

Ebb and Flow

(Physical Activity)

Birth and death rates of organisms in an ecosystem increase and decrease due to environmental factors such as disease or the lack of water, shelter, or food. Ebb and flow refers to variations in the birth and death rates of organisms in a specific ecosystem. To demonstrate how environmental factors affect the ebb and flow of populations, students can play the population game.

1. Designate one-third of the class to be an animal such as a coyote. Have these students wear **animal** name tags and stand in a line with their backs to the class.
2. Designate the rest of the class to be food, water, or shelter suppliers for the animals. Instruct the suppliers to form a line with their backs facing the animals. Have these students wear **supplier** name tags.
3. Give each student three factor cards; one labeled **food**, one labeled **water**, one labeled **shelter**.
4. Direct each animal to display one factor card to show what he or she needs to survive.
5. Direct each supplier to display one factor card to show what he or she can supply for the animals.
6. Signal students to turn and face each other.
7. Direct each animal to walk to a supplier that is showing a matching card.
8. When an animal cannot find a matching supplier, it "dies" and is out of the game.
9. Have students repeat the game until the supplier cards do not match any of the animal cards.
10. To illustrate the ebb and flow of the animal's population, invite students to play the game again.

Following the game, initiate a discussion about how environmental factors influence the ebb and flow of populations.

Biological Accumulation

(Art/Critical Thinking)

Biological accumulation occurs when organisms at the top of a food chain eat chemically contaminated organisms that are lower on the food chain. Because organisms higher on the food chain require more food, they consume more of the contaminated organisms. The contaminants accumulate in the bodies of these higher-order animals, causing health problems such as cancer or birth defects. Have students simulate this effect with the following demonstration.

1. Create a bar graph to record the levels of toxin each organism consumes. Fill in 25 squares on the first bar to represent 25 units of toxin. Twenty-five units is the amount of chemical one organism can safely ingest.
2. Sprinkle one-inch squares of paper on the floor. Each square represents five units of toxin found in one aquatic plant. Make a bar on the graph showing five units.
3. Choose 20 students to be herbivorous fish. Have each fish eat five aquatic plants by collecting five squares of paper. Make a bar on the graph showing the 25 toxic units one fish ingests from eating five plants.
4. Choose ten students to be carnivorous fish. Have each carnivorous fish eat two herbivorous fish by collecting all of their squares of paper. Make a bar on the graph showing the 50 toxic units one carnivorous fish ingests.
5. Choose five students to be aquatic birds. Have each aquatic bird eat two carnivorous fish by collecting all of their squares of paper. Make a bar on the graph showing the 100 toxic units one bird ingests.

Have students interpret the graph's results. Ask students the following questions: *Which organisms are below safe toxic levels? Which organisms are at the limit of safe toxic levels? Which organisms are above safe toxic levels?*

Adaptations

Living organisms adjust or change to help them survive the conditions particular to their ecosystem. Extreme environmental conditions, such as drought or flood, strongly influence the changes animals experience. These changes are known as **adaptations**. Some adaptations help organisms adjust to changes in water and temperature. Other adaptations help organisms obtain food from particular sources. Camouflage and certain body structures protect organisms from becoming prey. As an ecosystem changes, organisms either adapt or migrate in order to survive.

Specialized Structures

(Art)

Living organisms develop features that help them cope with certain conditions in their environment. Discuss examples of physical features plants and animals have that assist them in finding food, moving around, or staying alive. Provide students with a variety of art materials such as clay, pipe cleaners, tissue and construction paper, and craft sticks. Ask them to create an imaginary species of plant or animal with special adaptations for its environment. After they construct their organisms, ask them to share their new discoveries with the class. Encourage them to share the reasons they created certain features for the species.

Predator: Tiger

Adaptations	Functions
Sharp claws	to pull prey to the ground
Stripes	to hide from potential prey
Sharp teeth	to tear flesh
Strong legs	to

Predator and Prey

(Classifying/Interpreting)

Predators are animals that hunt other animals for food. Prey are animals that are hunted to become food for predators. Both have special adaptations that help them function in their roles. Have students study either a predator or prey and make an illustrated chart about the animal. Ask them to list the body structures and camouflage that enable the animal to be an effective predator or that protect it from becoming prey.

Adaptations

Bird Beaks

(Role Playing/Demonstration)

This activity demonstrates that adaptation has played a role in the development of bird beaks. Over time, bird beaks have adapted into shapes that help birds easily pick up food in their environments. Using familiar materials, have groups of four students role play different birds to determine which type of "beak" works best for gathering various kinds of "food." Instruct students in groups of four to follow these steps:

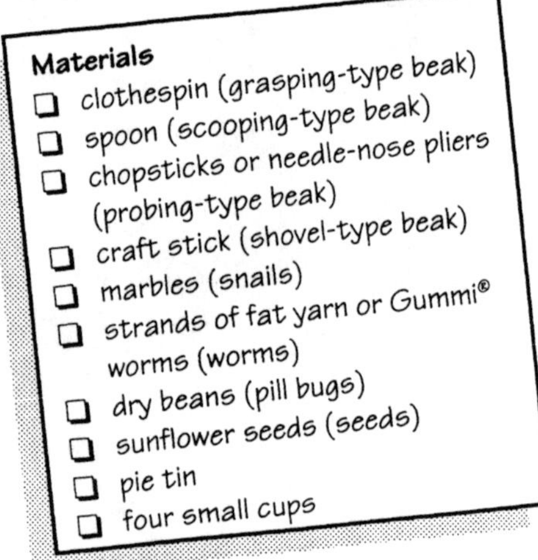

Materials

- ❑ clothespin (grasping-type beak)
- ❑ spoon (scooping-type beak)
- ❑ chopsticks or needle-nose pliers (probing-type beak)
- ❑ craft stick (shovel-type beak)
- ❑ marbles (snails)
- ❑ strands of fat yarn or Gummi® worms (worms)
- ❑ dry beans (pill bugs)
- ❑ sunflower seeds (seeds)
- ❑ pie tin
- ❑ four small cups

1. Choose a bird to role play and select a tool that represents its type of beak. Place a cup on the table in front of you to represent your stomach.
2. Pour "food" (marbles, yarn, Gummi® worms, beans, and seeds) in a pie tin. Choose one food to try to eat.
3. At a signal, pick up the "food" with your "beak" (spoon, pliers, clothespin, craft stick, chopsticks) and place it in your cup. If a piece of food is dropped, place it back into the pie tin. Collect only one piece of food, such as one seed, at a time.
4. After 30 seconds, count up the number of food pieces you ate. Record the information on a data table.

	# of "snails" eaten	# of "worms" eaten	# of "bugs" eaten	# of seeds eaten
Scooping-type beak				
Probing-type beak				
Shovel-type beak				
Grasping-type beak				

5. Return the food to the pie tin. Continue using the same beak and test for each of the remaining foods. After all foods have been tested, choose another beak and "eat" each food again.

Invite students to test all four bird beaks with each food by repeating steps two through five. Once students have tried each bird beak and have recorded their results, encourage groups to compare their findings and determine which beak is best suited for each food. Have each group present their findings to the class. As an extension, students can research different types of birds that have similar beaks to the ones they modeled to discover which food each bird eats in its habitat.

Extension Activities

Food Chains

(Physical Activity)

To illustrate how a food chain works, play a variation of tag called the *Food Chain Game.*

1. Set large outdoor boundaries to define an "eating area."
2. Designate another area to be a cemetery for "deceased" animals.
3. Choose two students to be eagles and six to be snakes. The rest of the students will be mice. Provide each mouse with a small plastic bag to represent its stomach. Create a way to distinguish the animals from one another, such as using name tags or hats.
4. Spread popcorn on the ground inside the eating area.
5. Signal the mice to run around the eating area for two minutes and "eat" the popcorn by placing it in their bags.
6. After two minutes, signal the snakes to chase the mice to "eat" them. A mouse is "eaten" when it is tapped by a snake. When a mouse is tapped, it gives the bag of popcorn to the snake and moves to the cemetery.
7. After two minutes, signal the eagles to go in and chase the snakes and remaining mice to "eat" them. When an eagle taps a snake or a mouse, the eagle takes the popcorn bag. All tapped snakes and mice move to the cemetery.
8. After two minutes, stop and count how many eagles got food, how many snakes and mice still have food, and how many snakes and mice have died.

Following the game, invite students to discuss how the simulation compares to a real food chain.

Storytelling

(Drama)

Have pairs of students use props and sound effects as they read aloud short fictional stories about the environment to the class such as *The Lorax, The Wump World*, and *The World That Jack Built.*

Big Al

(Art)

Read the story *Big Al* to illustrate examples of camouflage. Big Al, a lovable but ugly fish, tries to change himself so that he can fit in and have friends. In the end, his ugliness serves to save the day. Assign groups an area in the room to use as their habitat. Have each group design and decorate an animal to camouflage in that particular habitat. Gather all the animals in one place and ask students to determine in which habitat each animal belongs.

Water Ecology

Rain
One drop

Fell once
In France
Frolicked as it slid down the Eiffel Tower

Plummeted from
Prehistoric clouds
And rolled off the back of a dinosaur

Glided down
The cheek
Of a homeless man in New York
Shared his sorrow

Dropped in
On a rice paddy in China
Nourished it

Rain
The first time traveler.

—Brooks Osborne, 1994

"Rain" by Brooks Osborne. First appeared in *Merlyn's Pen: The National Magazines of Student Writing.*

Key Concepts for Water Ecology

The Water Cycle—

the circulation of the earth's water through evaporation, condensation, precipitation, and accumulation

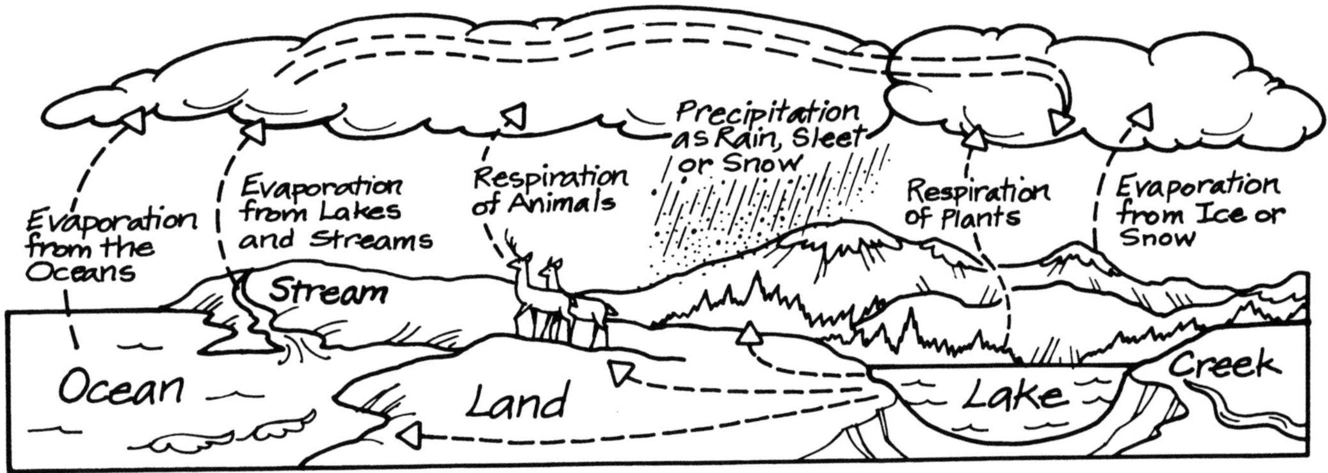

Water Pollution—

the contamination of water by synthetic and organic material

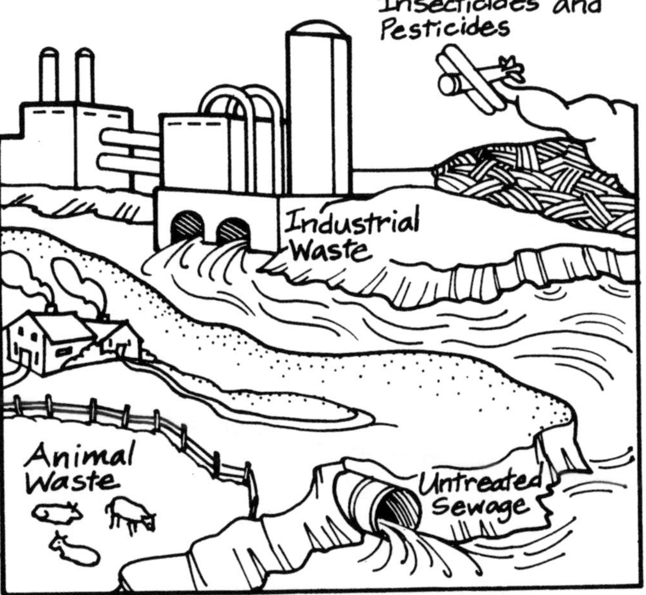

Water Conservation—

man's attempt to maintain, preserve, and restore water sources

Overhead reproducible

The Water Cycle

Water, as a solid, liquid, and gas, is found on and under the earth and in its atmosphere. Water travels through the **water cycle** to make its way from the earth, into the atmosphere, and back to the earth again. During the cycle, the sun's energy causes water from oceans and lakes to **evaporate** and rise into the atmosphere in the form of water vapor. The **water vapor** then **condenses** into cloud formations. When a cloud is too heavy to hold more water, water is released and falls back to the earth's surface as a liquid in the form of **precipitation**. This recycled water is used by living organisms or accumulates in storage places such as lakes, oceans, or water tables.

* *

Through the Cycle

(Science Demonstration)

Demonstrate the water cycle. Heat water in an electric frying pan and hold a cookie sheet filled with ice cubes about one foot above the water. The heat represents the heat from the sun. The pan represents the earth. The cookie sheet represents the cooler, upper atmosphere. As the water heats and evaporates, it becomes invisible and turns into water vapor. (In nature, the sun's heat causes evaporation without the boiling process.) The water vapor rises until it reaches the cool cookie sheet where it condenses into a liquid. As condensation continues, the water droplets on the bottom of the sheet get bigger and heavier until they begin to fall as precipitation. The precipitation falls back into the pan where the process begins again. Ask students the following questions: *Does all the water return to the pan? Why or why not? How is this experiment the same as or different from the real water cycle?* Invite students to record answers in their journals.

Life in the Water

(Observation/Comparison)

All living things need water for survival. Water is also a habitat for many creatures. Using small clear containers, help students obtain water samples from a variety of sources—oceans, streams, ponds, lakes, puddles, or runoff. Have them label each container with the name of the source and date. Ask students to use microscopes or magnifying glasses to observe these water samples for the presence of living organisms. In their science journals, students can sketch their observations with a description comparing the types of organisms they observe in the various sources.

The Water Cycle

Aquifers

(Science Demonstration/Observation)

Explain to students that in most areas of the world, if you dig deep enough, you'll find water because water seeps through the ground until it reaches a layer of rock that it can't get through. Water under the ground is called ground water. Ground water settles in an underground reservoir of loose sand and gravel called an aquifer. Wells can be drilled into aquifers to bring water to the surface. The point in the ground at which a drill reaches an aquifer is called the water table. Aquifers with a high water table are near the earth's surface. Aquifers with a low water table are far below the earth's surface. Have students make a model of an aquifer by following these steps:

1. Line a shallow box with a double layer of plastic wrap to represent the layer of rock that water cannot get through.
2. Fill the lined box ¾-full with sand.
3. Slowly pour water along the perimeter of the box until the sand is damp.
4. Dig a small hole with your finger in the center and line it with small stones to make a well.
5. Watch the water level in the well rise.
6. Add water each day to replenish the aquifer.

Have students keep a daily graph to show how much water is added to fill the aquifer. Invite students to discuss the results of the graph. Keep the aquifer intact to use in another demonstration later in this unit.

Salt Water into Fresh Water

(Science Experiment)

pages 22-23

Over 95 percent of the earth's water is in its oceans. Through the water cycle, the oceans' salt water is naturally distilled into fresh water which can be used by plants and animals. By designing a water distillation system in Experiment #2 on pages 22 and 23, students discover how salt water can be naturally transformed into fresh water. In the experiment, water vapor rises and condenses while salt remains in a large bowl. A small bowl creates a well where fresh water accumulates.

Name:

Salt Water into Fresh Water
Experiment 2

EXPERIMENT 2

Question: ***Can I make salt water into fresh water?***

Hypothesis: ______________________________

____

____

Procedure:

Materials
- ❑ saltwater solution
- ❑ large bowl
- ❑ small cup or bowl
- ❑ plastic wrap
- ❑ glass marbles
- ❑ rubber band

Step 1

Dip a finger in a saltwater solution (1 tsp. salt to 8 oz. of water). Taste the water from your finger.

Step 2

Place a small cup or bowl in the center of a larger bowl. Weigh it down with marbles.

Step 3

Carefully pour the saltwater solution into the larger bowl. Be sure not to get any salt water in the small bowl.

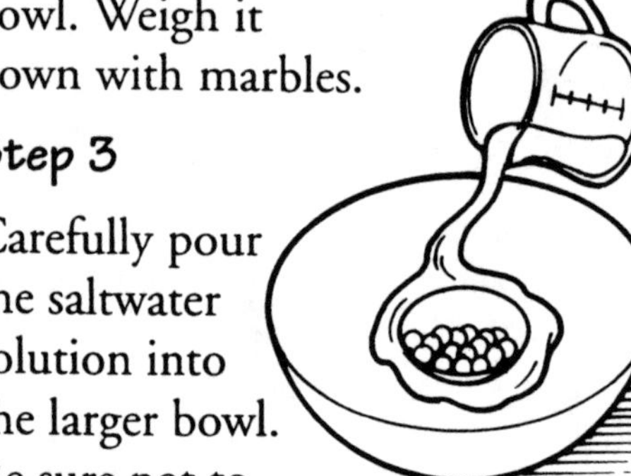

Step 4

Over the top of the large bowl, place plastic wrap loosely, allowing it to drape down into the center of the small bowl. Secure it with a rubber band around the rim.

Step 5

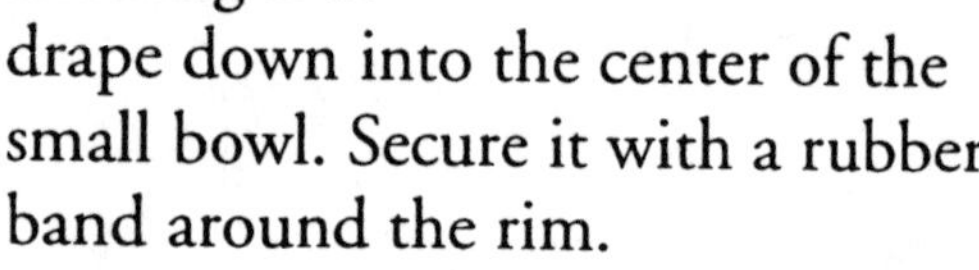

Place one marble in the center of the plastic wrap so it makes an inverted peak over the small cup inside.

Step 6

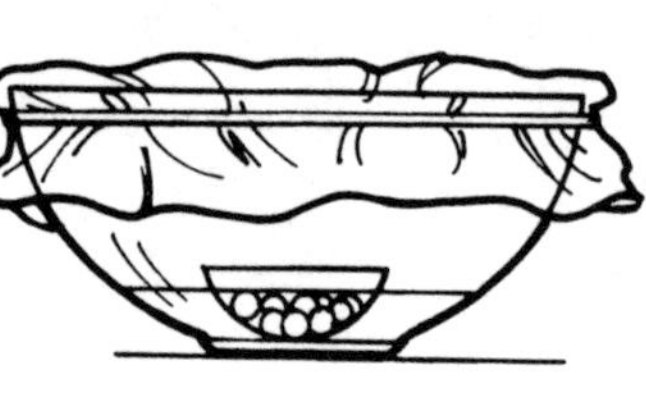

Place the distiller in a sunny location and observe it for at least 24 hours.

Step 7

Remove the plastic wrap. Taste the contents of the smaller cup.

Name:

Salt Water into Fresh Water
Experiment 2

EXPERIMENT 2

Results and Conclusions:

1. Does the water in the smaller cup taste salty? Explain what happened.

2. Do the results support your hypothesis? Why or why not?

3. Compare your results with your classmates' results. Did everyone have the same amount of water in the smaller cup? What factors could affect this?

4. How does this experiment demonstrate what happens during the water cycle in nature?

5. Based on this observation, list ways we can use this process to help us.

Science Challenge: Set up an experiment to test this question:

Can evaporation be used to remove water from oil?

Write your question, hypothesis, procedure, and materials list on another sheet of paper. Then test the hypothesis and record your conclusions.

Water Pollution

Water is a universal solvent used to dissolve and clean many substances. Some substances that mix with water may be organic such as waste or dirt. Other substances may be synthetic such as chemical **pesticides** or heavy metals. Some substances that mix with water **decompose** readily while others disintegrate slowly or never break down. Although water **purifies** itself through natural **filtration**, harmful substances can remain on the earth after water has evaporated. As the human population grows and engages in activities which produce harmful wastes, more dangerous substances find their way into our water sources.

****

Water Filter

(Science Demonstration/Observation)

As water seeps down through soil and rocks, it is cleaned and purified. Simulate this process by constructing a water filter. Using a flour sifter as the filter, layer absorbent cotton on top of the screening. Next layer coarse, clean sand. Place clean pebbles on the top layer. *Slowly* pour muddy water into the filter. Students can observe how the water is cleaned as it moves through the filter. Discuss what is happening and how the water is purified via the filter layers. Remind students not to drink the water because it may contain invisible bacteria or chemicals. Students can repeat this process several times and compare the water's clarity after each filtering. Have them document the steps of the procedure as well as the results.

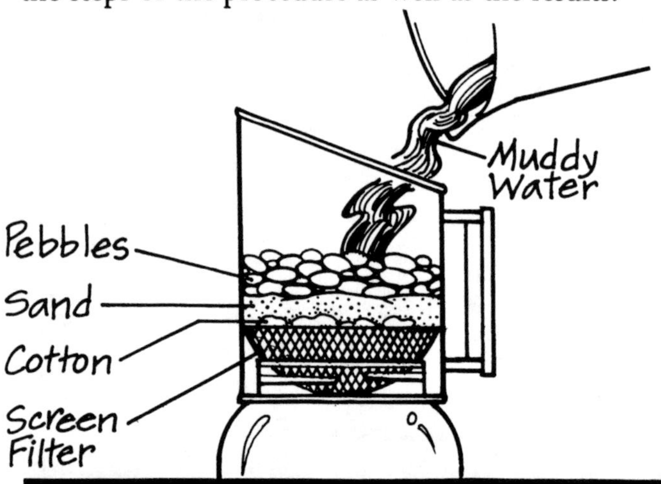

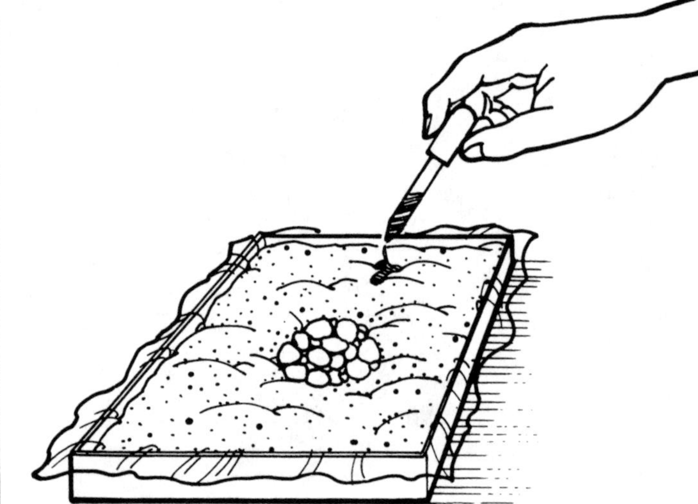

Water Pollutants

(Science Demonstration)

Even though soil, sand, and rocks naturally purify water, sometimes they carry pollutants that cannot be easily removed from the water cycle through natural filtration. Using the aquifer from page 21, introduce a pollutant into the ground water by placing a drop or two of food coloring on the outer edge of the sand. Continue to water as before. Have students observe the aquifer over time to see if the pollutant (food coloring) eventually disappears. The food coloring will not disappear because it cannot be removed naturally by the water cycle. Even when the water in the aquifer evaporates, color will remain in the sand near the well.

Water Pollution

In Hot Water

(Science Demonstration/ Writing)

Industries such as oil refining, steel production, and electricity production from power plants use water as a coolant. Because this water is not consumed, it is released back into the environment. Many times this waste water is returned at a temperature higher than normal for that water environment, causing thermal pollution. Thermal pollution changes the environment for aquatic life; a temperature change can kill aquatic plants or interfere with fish reproduction, growth, and food supply. To demonstrate the effects of thermal pollution on plant life, perform this demonstration:

1. Obtain two Anacharis plants from an aquarium supply store.
2. Obtain two small goldfish bowls. Label the bowls *hot* and *cold.* Place the bowls in two deep pails or tubs to catch runoff. Fill the goldfish bowls with room-temperature distilled water.
3. Place one plant in each bowl. Each day, have students add distilled water at room temperature to the cold bowl and hot distilled water to the hot bowl.
4. Ask students to observe the two plants for one to two weeks, watching for differences. The Anacharis in the hot bowl will become brittle and begin to die.
5. Have students record their findings in science journals and answer these questions: *What are the consequences of thermal pollution for plants and animals? What steps could be taken to eliminate thermal pollution and restore plant and animal life?*

Phosphates

(Observation/ Writing)

Although soaps and detergents are useful in our homes, they can be harmful to the environment. They often contain phosphates which can affect organisms in the water. To observe how phosphates affect living organisms, ask students to follow these procedures:

1. In a small container, make a detergent solution by mixing eight spoonfuls of water with one spoonful of phosphate detergent.
2. Collect eight ounces of stagnant water from outside. For best results use pond water.
3. Put half of the water into a small jar and label it *control sample.*
4. Put the other half of the water in another small jar. Add a few drops of the detergent solution. Label this jar *phosphate sample.*
5. Put both jars in a sunny window. Observe over a few days.

The phosphates overfeed the algae that are in the water. The algae use up the oxygen in the water and begin to grow rapidly. The algae block sunlight by forming a mat on the water's surface. In a natural setting, the blockage of sunlight can inhibit aquatic plant growth and harm aquatic animal life. Ask students to respond to the following questions in their journals: *What did you observe? How might phosphates affect living organisms in a water environment?*

Water Pollution

Pollution Solutions

(Writing/Critical Thinking)

Post a large classroom chart and invite students to brainstorm a list of pollution problems. Record their ideas in a vertical column. In a second column, chart possible solutions. In a third column, list the pros and cons of each solution. Encourage students to think of ways the solutions can be implemented. As an extension of this activity, students can create personal charts with a fourth column that lists ways they can help alleviate water pollution.

POLLUTION SOLUTIONS			
Problem	**Possible Solution**	**Pros and Cons**	**How I Can Help**
contaminated lakes	large submarines	**pro** - fast clean-up **con** - expensive	pick up trash at the beach
algae-filled pond	algae vacuum	**pro** - fast clean-up **con** - short-term solution	use non-phosphate detergent
dying fish in a pond near a factory that pumps hot water	restock the pond	**pro** - great fishing **con** - fish will still die	write letters to the EPA and my congress-person

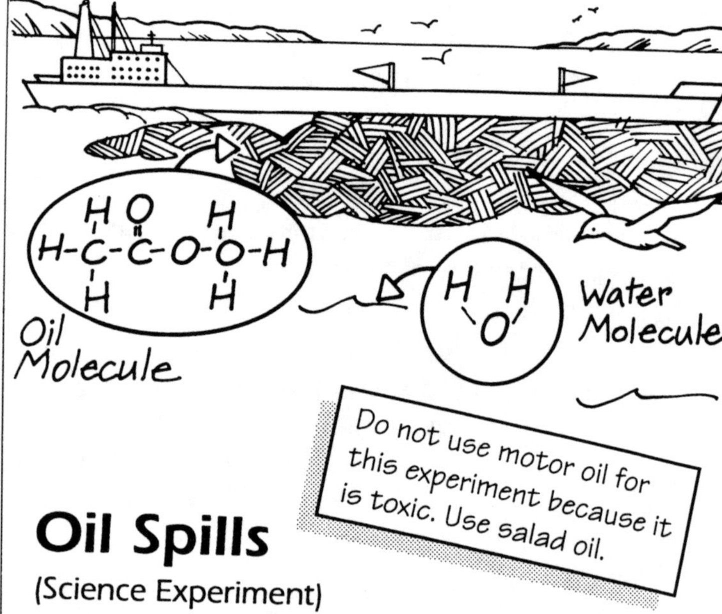

Do not use motor oil for this experiment because it is toxic. Use salad oil.

Oil Spills

(Science Experiment)

pages 27-28

Oil tankers deliver the oil needed for cars, homes, and businesses. When these tankers leak or spill oil, the oil forms a slick on the water's surface. An oil slick forms because oil emulsifies and naturally stays separate from water. Since oil is lighter and less dense than water, it floats on top. An oil slick dissipates in a variety of ways. Up to half of the slick evaporates quickly. The remaining oil is broken up into small particles and remains suspended in the water. The sun helps break down some of the small particles. After about a month, marine organisms and bacteria biodegrade more of the remaining oil. The rest of the oil slick forms tar balls which remain in the environment. An oil slick is devastating to an ecosystem for various reasons—it prevents sunlight from reaching below the surface of water; it causes animals such as birds and otters to drown; it allows toxins to contaminate the water supply. When oil eventually washes ashore, it begins to contaminate the soil. In Experiment #3 on pages 27 and 28, students simulate an oil spill and test materials to see how well they help clean up a spill.

When the experiment is complete, have students decide the best way to dispose of the materials they used to clean up their oil spill.

Name:

Oil Spills
Experiment 3

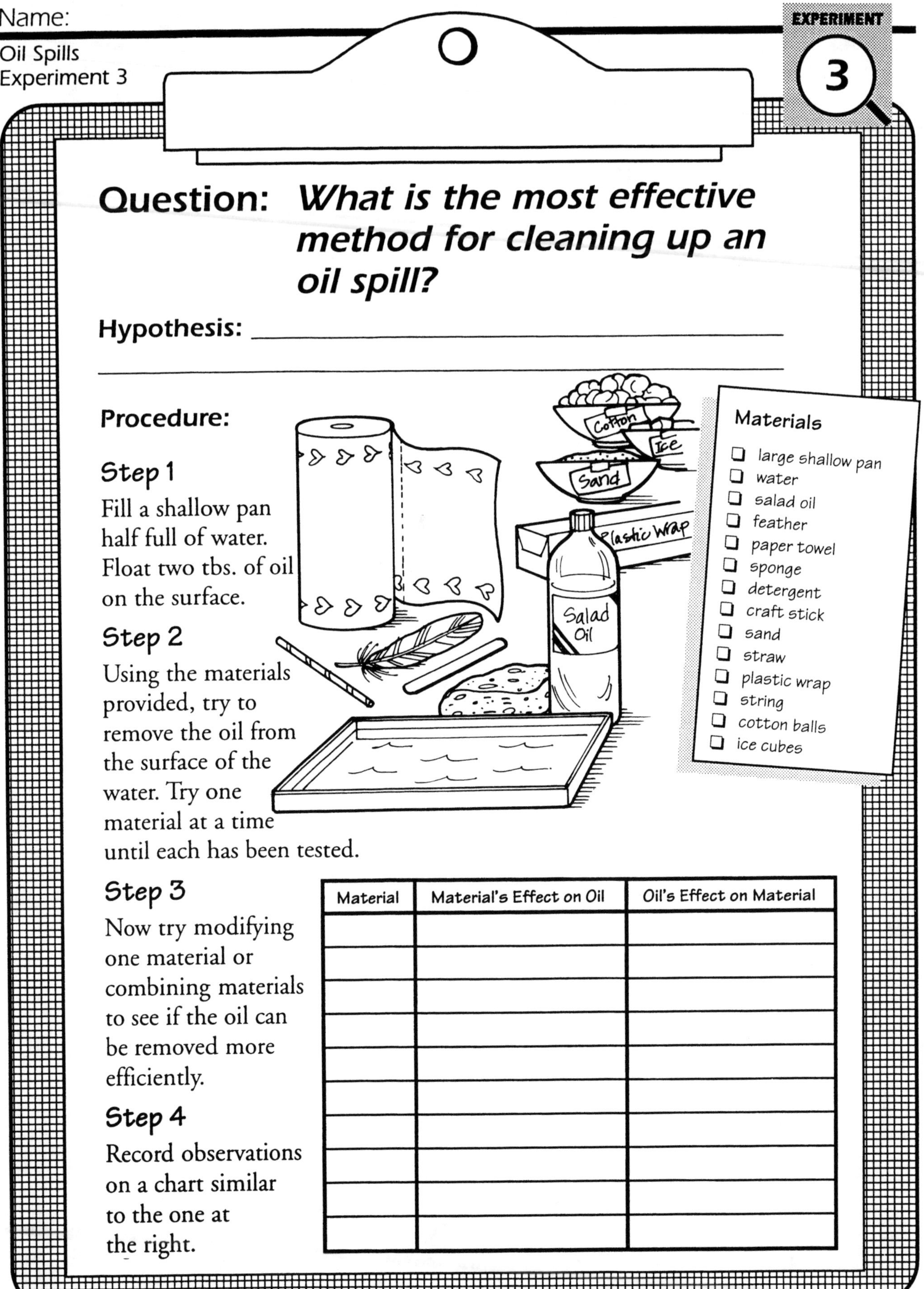

Question: ***What is the most effective method for cleaning up an oil spill?***

Hypothesis: ______________________________

____

Procedure:

Materials

- ❑ large shallow pan
- ❑ water
- ❑ salad oil
- ❑ feather
- ❑ paper towel
- ❑ sponge
- ❑ detergent
- ❑ craft stick
- ❑ sand
- ❑ straw
- ❑ plastic wrap
- ❑ string
- ❑ cotton balls
- ❑ ice cubes

Step 1

Fill a shallow pan half full of water. Float two tbs. of oil on the surface.

Step 2

Using the materials provided, try to remove the oil from the surface of the water. Try one material at a time until each has been tested.

Step 3

Now try modifying one material or combining materials to see if the oil can be removed more efficiently.

Step 4

Record observations on a chart similar to the one at the right.

Material	Material's Effect on Oil	Oil's Effect on Material

EXPERIMENT
3

Results and Conclusions:

1. Do your results support your hypothesis? Why or why not? ______

2. What material seems to work best? Do you think this material could be used to clean up large oil spills? Why or why not? ______

3. Did you make any modifications to your materials or combine materials? Describe what you did and its effect. ______

4. Compare your experiment with an actual oil spill. What knowledge have you gained about the challenges of cleaning up an oil spill?

5. Based on your observations, what other materials could you try to clean up this spill? What effect do you believe they would have?

Science Challenge: Set up an experiment to test this question:

Can I use a combination of materials to invent a tool that will help remove oil from water?

Write your question, hypothesis, procedure, and materials list on another sheet of paper. Then test the hypothesis and record your conclusions.

Water Conservation

People continue to harness water and mold it to fit their needs, sometimes without regard for the environment. Because water is a limited **resource** and the earth cannot sustain life without it, steps must be taken to **conserve** it. Water conservation involves a conscious attempt to **maintain, restore,** and **preserve** water sources. Cleaning up lakes and streams helps rebuild water ecosystems. The use of nonpolluting cleaning supplies prevents further pollution of water sources. Water-restricting devices in faucets help limit water usage. With continued conservation, people can preserve the earth's water supply.

* *

Xeriscaping

(Research/Writing)

Xeriscaping is landscaping with native plants. The use of native plants rather than lawn or high-maintenance plants can conserve water. Have students use resource materials or contact local arboretums, horticulture societies, and nurseries to research native plants in their area. Students can use this information to draw aerial views of their yards at home or other grassy areas, replacing the lawn with native plants. Encourage students to gather additional information on the cost of the replacement materials and to calculate the total cost of the project. Students can prepare budgets to support their research.

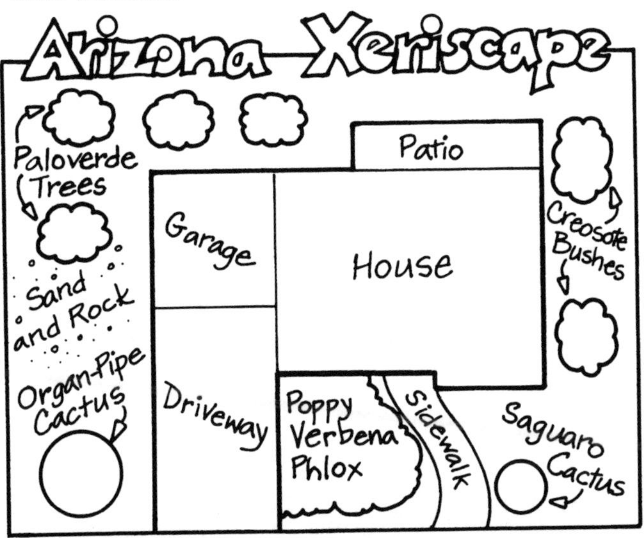

Threatened Waterways

(Research/Writing)

In the 1960s and 1970s people became concerned about Lake Erie, the smallest and most shallow of the Great Lakes. Dead fish and algae washed up on shore; blue-green algae covered the surface of the water. Fishing and swimming were banned due to possible health hazards. Have students work in groups to study a threatened water source such as a lake, river, wetland, or bay. Encourage them to discover what caused the pollution and what steps have been taken to restore the water source. Have groups make two posters depicting two views of their water source—a "before" scene of the polluted area and a second "after-the-clean-up" scene. Invite each group to make an oral report describing their posters.

Water Conservation

Water Surveys

(Research/Critical Thinking)

For one week, have students keep track of all the ways their family uses water at home. List these uses on a chart. Next to each use, have students record the priority of this water use. For example, water for a baby sister's formula is necessary and therefore a high priority. Hosing down the driveway is a low priority. In a third column, have students list ways to reduce the amount of water needed for that use or to reduce the amount of pollutants added to the water for that use.

Water Use	Priority	Ways to Reduce Water Use
a drink of water	high	Don't run the water before filling the glass.
hosing the driveway	low	Sweep.
watering plants	medium	Plant drought-tolerant plants.

Be Active!

(Field Trip/Research)

Plan a field trip to a local water treatment facility or a pumping station for a town well. Before the trip, have students study concepts such as coagulation, filtration, disinfection, and water softening. Invite students to write a series of questions to ask facility employees. On the day of the trip, have students ask employees their questions and record the answers in their journals.

After the field trip, have students contact Adopt-a-Stream, an organization that helps organize volunteer programs to clean up water sources and monitor water quality. To obtain information write to:

Adopt-a-Stream
P. O. Box 435
Pittsford, NY 14534-0435

Extension Activities

A River Ran Wild

(Art/Research)

Share the story *A River Ran Wild* by Lynne Cherry with your students. Have them draw or paint a mural of the winding Nashua River in its natural state. Have each student choose an animal from page 8 of the book to study. Ask students to illustrate and cut out their animal. On an index card, have students write the name of the animal and the ways it is dependent upon the river. Add the animal cutouts and cards to the mural. Around the border of the mural, have students draw pictures of activities that endanger the river. Cover each picture with the universal "no" symbol (a circle with a diagonal line through it).

Letter Writing

(Writing)

Review and practice letter-writing skills by writing for earth-friendly products. Invite the class to compose one business letter together requesting information about available products. After receiving information, encourage students to write letters to friends persuading them to change to earth-friendly products. Contact The Ecology Center for information about earth-friendly products at:

2530 San Paulo Avenue
Berkeley, CA 94702

Geography, Climate, and Water

(Research/Writing)

Challenge students to do further research on biomes to discover the relationships between geography and rainfall. Have students find out which biomes have substantial, moderate, or little water and rainfall. After students complete this research, challenge them to find out how plants and animals are specially suited to obtain the water they need to survive in a particular biome.

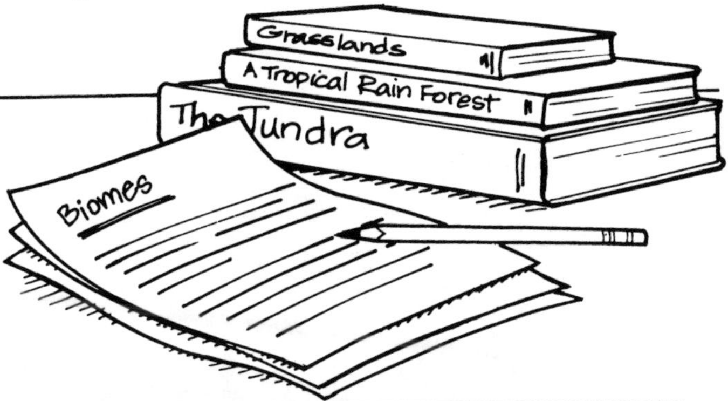

Air Ecology

Laying on a field of grass
under a sea of sky,
a spring breeze
catches my breath,
and somehow
it seems like a
winter night star
shines in me.

—*Bette McIntire,*
1994

Key Concepts for Air Ecology

The Oxygen/ Carbon Cycle—

the circulation of carbon and oxygen during photosynthesis and respiration

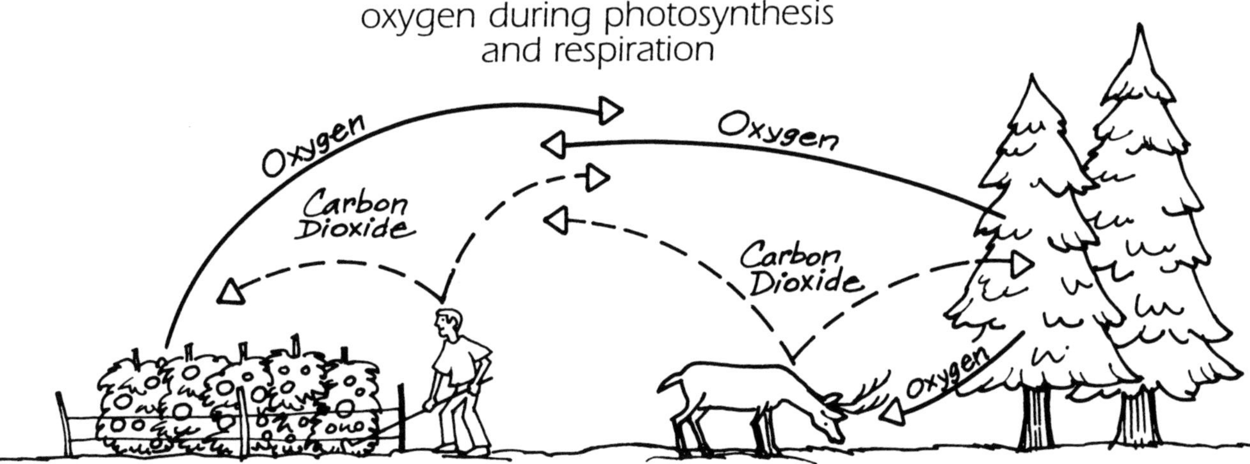

Air Pollution and Depletion—

the discharge of contaminants into the air damaging its quality and decreasing its quantity

Air Quality Control—

the maintenance and restoration of the air to a level that sustains the needs of the environment

Overhead reproducible

The Oxygen/ Carbon Cycle

The earth's atmosphere consists of 78% **nitrogen (N_2)**, 21% **oxygen (O_2)**, 1% **carbon dioxide (CO_2)**, and other trace gases. Carbon dioxide (CO_2) is released into the air by animals as they exhale. Plants take in CO_2 during **photosynthesis**. As a result of photosynthesis, plants produce and release oxygen. This oxygen is used by living things for **respiration**. The circulation of O_2 and CO_2 creates the oxygen/carbon cycle. Air and the constant recycling of its gases through the oxygen/carbon cycle allows organisms to survive.

****

Respiration

(Science Demonstration/Observation)

During respiration, animals inhale air containing oxygen and exhale air containing carbon dioxide. An animal uses oxygen for basic processes such as digestion and movement. During these processes, carbon dioxide is released and is exhaled from the animal's body. Conduct the following demonstration to observe the carbon dioxide we exhale. In a clear plastic container, add six drops of Bromothymol Blue to 1/3 cup of water. (Bromothymol Blue, available at aquarium supply stores, is a solution that reacts to acid in water by changing color.) Exhale into a straw to create bubbles. The exhaled CO_2 changes the water from blue to green and to yellow because CO_2 makes a weak acid solution. A blue color means the water has an adequate amount of oxygen and yellow shows the presence of acid in water. Invite students to diagram their observations in their science journals.

Photosynthesis

(Science Experiment)

Plants make their own food through photosynthesis by converting solar energy into chemical energy. Review with students the steps of photosynthesis. Discuss the consequences of a lack of oxygen for plants and animals.

Photosynthesis:

1. Plants take in water through their roots and CO_2 through the pores on their leaves.
2. When the leaves are exposed to light, CO_2 and water are converted to energy. Some of that energy is absorbed by chlorophyll in the leaves.
3. The energy in the chlorophyll acts as a catalyst to start chemical reactions.
4. Water in the plant is split into hydrogen and oxygen.
5. The hydrogen combines with carbon dioxide to store chemical energy by making a simple sugar called glucose.
6. During photosynthesis excess oxygen is released into the air.

In Experiment #4 on pages 35 and 36, students observe how the absence of light halts photosynthesis, causing carbon dioxide to build up and no oxygen to be released. Students will note the presence of carbon dioxide through the use of Bromothymol Blue.

EXPERIMENT 4

Question: *Without light, can carbon dioxide be used and converted to complete the process of photosynthesis?*

Hypothesis: ____

____

____

Procedure:

Step 1

Add 100 ml or ½ cup of water to each plastic cup.

Label one cup *bottled water* and the other *plant water.*

Step 2

Add six drops of Bromothymol Blue to each cup of water.

Step 3

Place several strands of Anacharis into the cup labeled *plant water*.

Step 4

Cover the cups with plastic wrap and secure the wrap with a rubber band.

Step 5

Place the cups in a dark location overnight and check the water in 24 hours. Complete questions one and two on your lab sheet.

Step 6

Place the cups in a sunny location for 30–45 minutes. Observe. Complete questions three through six on your lab sheet.

Materials

- ❑ two clear plastic cups
- ❑ plastic wrap
- ❑ two rubber bands
- ❑ one cup or 200 ml of bottled water
- ❑ Anacharis
- ❑ Bromothymol Blue

Anacharis and Bromothymol Blue are available at aquarium supply stores.

Bottled Water

Plant Water

EXPERIMENT 4

Results and Conclusions:

1. After 24 hours, observe the water in your cups. Explain what has happened.

2. Compare your cups of water to those of your classmates. Are they the same? Why or why not?

3. Place the cups in a sunny location for 30–45 minutes. What do you observe?

4. What effect did the sun have on the water?

5. How would the information you learned help you maintain a freshwater environment for fish?

Science Challenge: Set up an experiment to test this question:

Can additional aquatic plants convert more carbon dioxide and therefore increase the amount of oxygen in water?

Write your question, hypothesis, procedure, and materials list on another sheet of paper. Then test the hypothesis and record your conclusions.

Air Pollution and Depletion

Air **pollution** occurs when contaminating particles are introduced into the air. Fumes from **fossil fuels**—oil, coal, and other synthetic substances—burden the atmosphere with contaminants. Although the water cycle continually cleans the air of these particles, it is unable to clean contaminants as quickly as they are added. Air pollution also adds excess carbon dioxide to the air which may contribute to global warming. Without future action, air pollution could severely threaten the earth's ecosystems.

* *

Holes in the Ozone

(Research/Problem Solving)

The ozone is a layer of a form of oxygen (O_3) just inside the stratosphere, 10 to 30 miles above the earth. It protects the earth against harmful rays from the sun. Scientists speculate that if this layer is depleted, the harmful rays will penetrate the earth's atmosphere and cause health problems. Chemicals called chlorofluorocarbons (CFCs) damage the ozone. These chemicals are used as coolants in refrigerators, car air conditioners, and spray cans. Recent legislation requires manufacturers to use materials that pollute less than CFCs. Have students contact car and appliance manufacturers to ask specific questions about how they have adapted their products to comply with new regulations. Students can present oral reports with their results to the class.

Can You See It?

(Physical Activity/Critical Thinking)

Air pollution is not always visible. Have students spread a very thin layer of petroleum jelly on strips cut from plastic milk jugs. Ask students to hang the strips in a variety of locations where they will not be disturbed. Leave the strips hanging for a week; then have students check them for any particles that have adhered to the petroleum jelly. Discuss possible sources of these particles. Remind students that these particles are in the air they breathe. Challenge students to think of ways they can filter the air. Encourage them to collaborate and create visual examples of their ideas, using a variety of materials such as pipe cleaners, construction paper, and fabric. Display the inventions for students and visitors to see.

Air Pollution and Depletion

The Greenhouse Effect

(Art/Demonstration)

One theory of global warming, the greenhouse effect, states that too much carbon dioxide in the air may raise the temperature of the earth. The amount of carbon dioxide entering the air increases when fossil fuels are burned. The excess carbon dioxide cannot be used by plants during photosynthesis. This excess carbon dioxide absorbs heat energy from the sun, keeping the heat near the surface of the earth. This heat raises the earth's temperature. A temperature increase of two or three degrees in the average temperature of the earth could trigger disasters such as the melting of the polar ice caps, which would cause the oceans to rise and flood the earth's coastlines. To demonstrate the theory of the greenhouse effect, have groups make clay models of a coastline (in an aluminum tray) which include riverbeds, inland and coastal cities, mountains, and lakes. Have students follow these directions:

1. Fill the "lakes" with water.
2. Place a block of ice, frozen in a paper cup, on the tallest point of the model.
3. Cover the entire model with plastic wrap to represent the layer of carbon dioxide around the earth.
4. Leave the model in a warm, but not hot, location until the ice melts.
5. Observe the results.

Encourage students to speculate where the ice went. Have them write stories about what would happen to the environment if the polar ice caps melted.

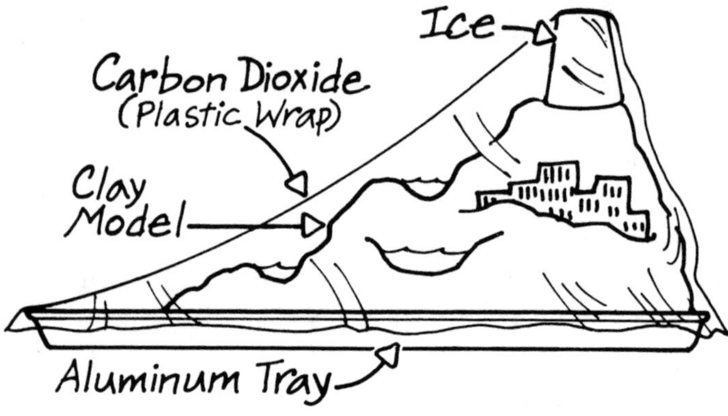

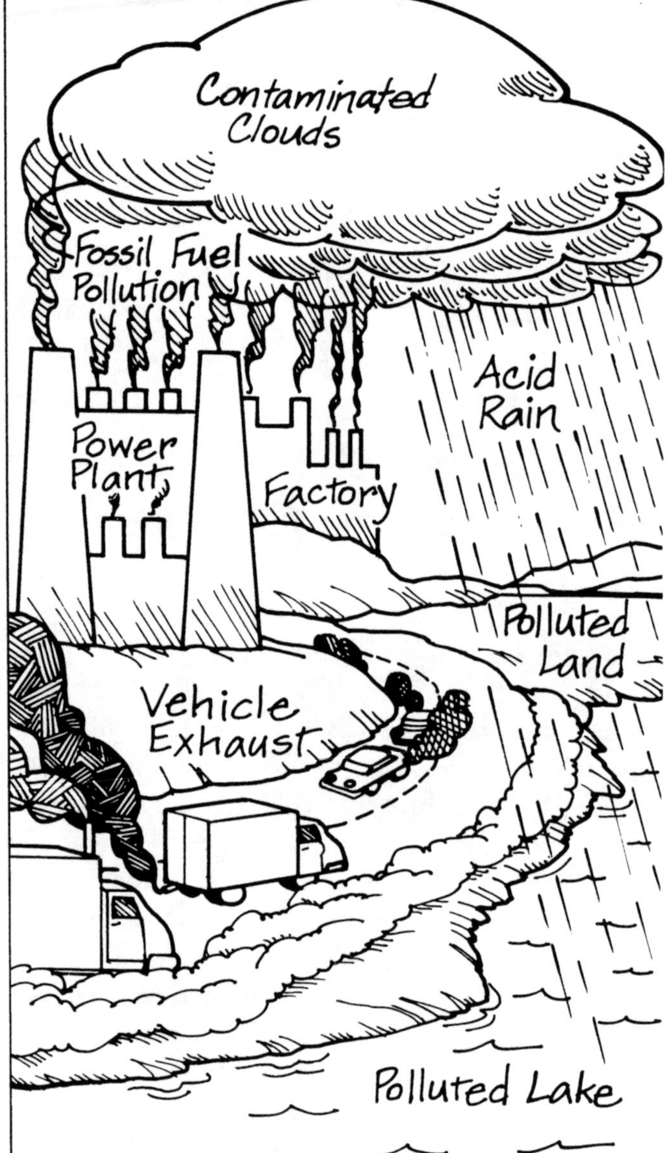

Acid Rain

(Science Experiment)

When factories burn fossil fuels and cars release gases into the atmosphere, pollutants mix with moisture in the air to create clouds and eventually fall to the earth as acid rain. When lakes, rivers, soil, and plant life are exposed to acid rain, the acid affects the health of living organisms. In Experiment #5 on pages 39 and 40, students show how acid rain affects plant growth.

EXPERIMENT 5

Question: *Can plants tolerate the effects of acid rain?*

Hypothesis: ______________________________

Procedure:

Materials

- ❑ vinegar
- ❑ tap water
- ❑ two large containers
- ❑ two small plants of the same type
- ❑ measuring spoon
- ❑ measuring cup

Step 1

Prepare two containers of watering solution—one plain tap water and one acid water. Make the acid water by adding one tablespoon of vinegar to one cup of tap water.

Step 2

Label one plant *control* and the other *acid.*

Step 3

Place both plants in a location with adequate light and the appropriate temperature. Water the control plant with regular water and the other plant with acid water. Water each plant with the same amount.

Step 4

Make daily observations about the plant growth and soil in both containers. Water when necessary with the appropriate solution (approximately one to two times per week). Keep a record of your observations in your science journal for two weeks.

EXPERIMENT 5

Results and Conclusions:

1. Describe the effects of the acid water on your plants.

2. Describe any observable changes to the soil.

3. Do the results of your experiment support your hypothesis? Why or why not?

4. Compare your results with those of your classmates. How are the results similar? How are they different?

5. What conclusions can you make about how acid rain affects soil quality and plant growth?

6. What are some ways you could lessen the acid in the acid rain solution? How would this affect the plant?

Science Challenge: Set up an experiment to test this question:

Can the effects of acid rain be corrected by adding lime to soil?

Lime can be obtained at a nursery or garden center.

Write your question, hypothesis, procedure, and materials list on another sheet of paper. Then test the hypothesis and record your conclusions.

Air Quality Control

The burning of fossil fuels is the greatest cause of air pollution. We burn fossil fuels to drive vehicles, power factories, make electricity, and heat and cool buildings. In order to **restore** and **protect** our **air quality**, we must reduce consumption of fossil fuels. The conservative consumption, manufacturing, and disposal of chemicals and plastics can also help improve air quality. Through regulation and responsible behavior by industry and consumers, we can greatly improve the quality of the air we breathe.

* *

Reducing Fossil Fuel Dependence

(Problem Solving)

Automobiles burn half of the oil in the United States. By reducing the amount of time we spend in cars and the number of cars on the road, we can reduce air pollution. Have students keep a record of every trip in a car or bus for a week. At the end of the week, compile a class chart. Encourage students to brainstorm ways to reduce car use, such as planning trips for more than one purpose, using alternative methods of transportation, or carpooling. Students can write letters to family members and friends to explain how they can improve the air quality by reducing the amount of time spent in a car. Have students ask for help by listing ways to accomplish their goal.

Activity	Transportation	Alternatives
getting to school	car	ride a bike
going to soccer game	car	carpool with friends
going to the movies	bus	walk

Energy and the Air

(Writing/Critical Thinking)

It is not commonly known that the energy used for homes and businesses can affect the air. The burning of fossil fuels powers generators that produce electricity. By using less energy, people can reduce the need for fossil fuels. Have groups of students write letters to local power and gas companies requesting information on how they can reduce energy usage at home or school. Send a few letters to each company. Using this information, the students can prepare energy conservation plans. Have students write reports and draw posters to present their findings and plans to the class.

Air Quality Control

Plant a Tree

(Critical Thinking/Community Involvement)

Trees provide shelter for animals, help keep air fresh, and add beauty to the world. Encourage students to plant trees at school and in the local community. Students can develop a plan for fund raising, acquiring materials, and planting the trees. Have students use their plans to write persuasive letters to their principal, the school board, local nurseries, and community organizations such as the Chamber of Commerce. Once permission is obtained to plant trees, students can implement their plans. Photograph the planning and planting for a class book.

How Clean Is It?

(Demonstration)

Ask students, *Do you think the air in our classroom is clean?* Discuss what would cause it to be dirty. To demonstrate how to check the air for cleanliness, bring in a vacuum cleaner with a hose attachment. Place a piece of white facial tissue over the end of the hose and secure the tissue with a rubber band. Turn the vacuum cleaner on and wave the hose in the air for one minute. Remove the tissue and invite students to examine the tissue for particles by using a magnifying glass or microscope. Discuss possible sources of the particles and write student responses on a chart. Next bring in the air filter from your car for examination, or ask the school custodian for the air filter from your classroom. Discuss the sources of pollutants for these filters and document them on the chart. Encourage students to respond to the following questions: *Why would it be necessary to have these filters? What sources of pollution are these filters trapping? How could the filters be used differently to alleviate other air pollution problems?* Students can record responses in their science journals.

Extension Activities

Literature

(Critical Thinking)

In the ecological tale *The Wump World,* a clean and peaceful planet is threatened by the arrival of aliens called the Pollutions. The Pollutions have come to Wump World because their polluted planet is no longer inhabitable. Have students create a chart listing characteristics of the Wump World environment before and after the arrival of the Pollutions.

Before the Pollutions	After the Pollutions
land covered with vegetation	land covered with concrete
open countryside	crowded cities
clean lakes and streams	polluted waste water

In the story, the Interstellar Council of Inhabitable Planets is concerned with the behavior of the Pollutions as they move from planet to planet, polluting and destroying at will. Have students assume the role of an Interstellar Council member. In pairs, students can record their dialogue as council members. Have students discuss how to deal with the Pollutions, how to persuade them to stop destroying other planets, and how the Pollutions should live when they return home. Ask student pairs to share their dialogues.

Fuel Consumption Worksheet

(Math)

Have students use the reproducible on page 44 to calculate automobile fuel consumption. The objective is to have students realize that combining car trips saves time, gas, and money.

Use of Satire

(Writing)

Explain to students that satire is the use of sarcasm, irony, or wit to describe foolish behavior. Encourage students to consider how *The Wump World* is a satire of human behavior. Ask: *How are we like the Pollutions? In what ways can our behavior be considered different?* Encourage students to write satirical stories illustrating an ecological problem and/or solution.

Name:

Fuel Consumption Worksheet

Making several car trips to many destinations can use more fuel than making one long trip to all of the destinations. Answer the following questions to calculate how much fuel a car consumes as it travels to different destinations.

Fuel Consumption

1. Draw a diagram illustrating your home and three routes to three different destinations. Label the distances in approximate miles (see example).

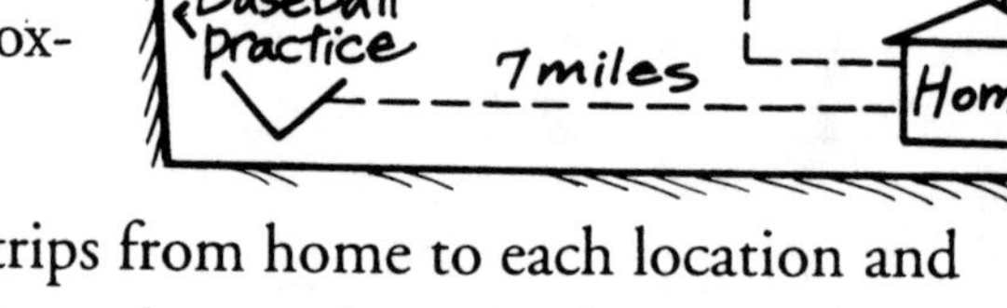

2. Calculate three individual trips from home to each location and back. Record the mileage for each round-trip in the spaces below:

Example: home to school to home = 10 miles

home to ____________________ to home = ______

home to ____________________ to home = ______

home to ____________________ to home = ______

3. Find the total distance traveled by adding together all three round-trip distances. Write your answer here. ____________________

4. On lined paper, combine visits into one trip to make the distance traveled shorter.

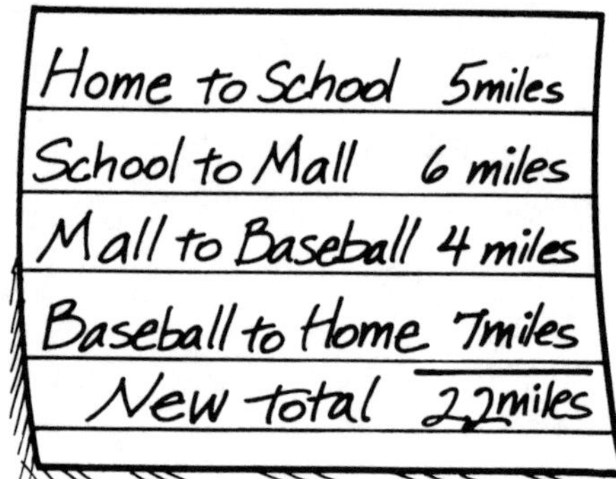

5. Calculate how much fuel the car would consume on each trip if the car got 20 miles/kilometers to the gallon/liter.

Use the following formula.

distance traveled ÷ 20 = how much gas would be used per gallon/liter

Use the following formula to calculate how much the gas cost.

gas used × cost of gas per gallon/liter = total cost

Soil and Land Ecology

We love this land as a newborn
loves its mother's heartbeat.
If we sell you our land, care for
it as we have cared for it.
Hold in your mind the memory
of the land as it is when you receive it.
Preserve the land and the air and
the rivers for your children's children
and love it as we have loved it.

—Susan Jeffers,
1991

Key Concepts for Soil and Land Ecology

The Nitrogen Cycle—

the transfer of nitrogen from the atmosphere into the soil and along the food chain

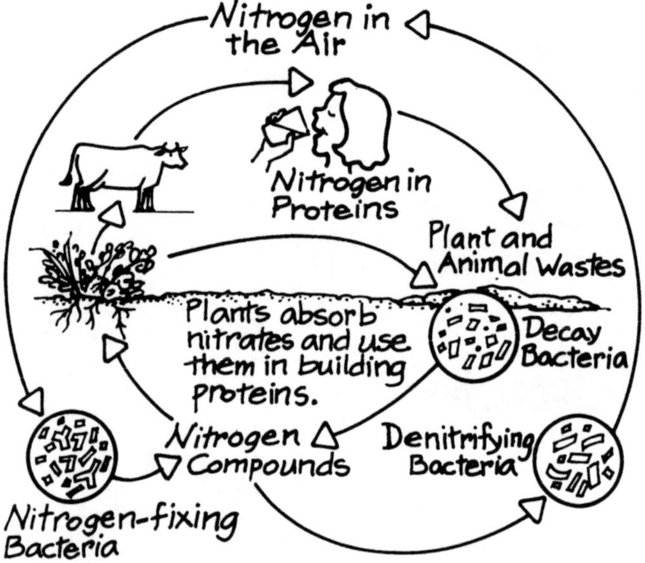

Soil and Land Exploitation—

abuse of the earth's soil and its resources due to human consumption and waste

Soil and Land Conservation—

man's attempt to maintain, preserve, and replenish the earth's soil and its resources

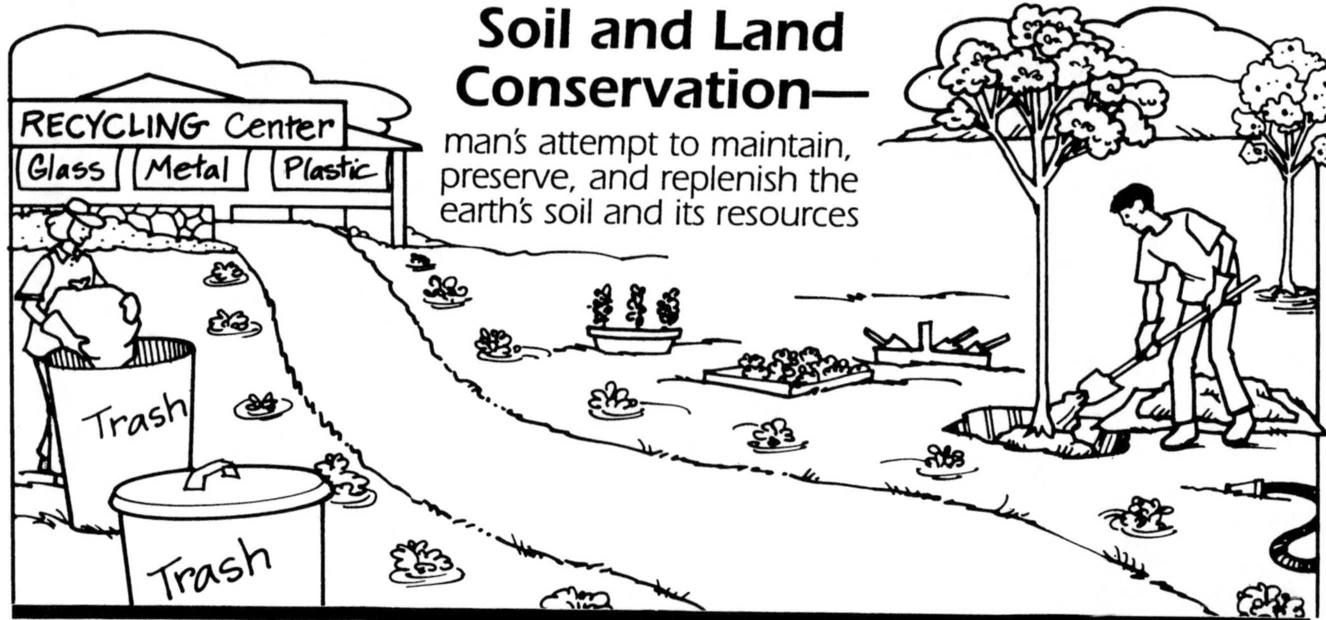

Overhead reproducible

The Nitrogen Cycle

The **nitrogen cycle** ensures rich soil, a stable atmosphere, and life for plants and animals. **Nitrogen** circulates through the atmosphere as an **element** and through the earth as a **compound**. Plants absorb nitrogen compounds from the soil. Animals then eat the nitrogen-rich plants, passing nitrogen along the food chain. When plants and animals die, they **decompose** and react with nitrifying bacteria to make nitrogen compounds in the soil that are used by plants. To regulate the amount of nitrogen in the soil, denitrifying bacteria change some of the nitrogen compounds into a gas while nitrogen-fixing bacteria change some of the gas into nitrogen compounds.

****

Nitrogen and the Food Chain

(Critical Thinking/Art)

All living things need nitrogen to live. Nitrogen is a vital element in the protein molecules found in protoplasm, the living substance in all plants and animals. People and animals get protein from eating other animals and plants, thereby transferring nitrogen along the food chain. Have students choose a food chain to study. Ask students to trace nitrogen through the food chain. Have students demonstrate nitrogen entering a plant from the atmosphere, leaving that plant, and continuing through the food chain. Inform students that nitrogen is released back into the environment when organisms dispose of waste or decompose. Encourage students to show the nitrogen cycle in a food chain through posters, dioramas, or skits.

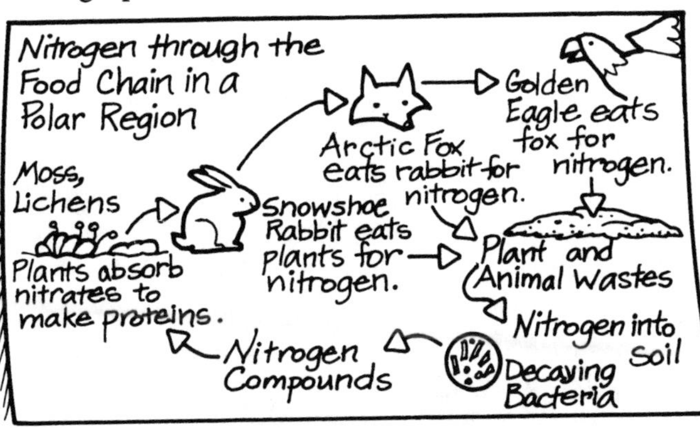

Compost It!

(Science Activity/Critical Thinking)

Soil needs organic matter, nutrient-rich decomposed plant and animal tissue, for plant growth. To create organic matter, have students make a mini compost pile. Invite students to use a variety of containers to layer fruit and vegetable waste with potting soil, leaves, and grass clippings. Have students cover the containers and place them in a sunny, outdoor location. Ask students to shake the containers every other day to add air to the mixture. If the weather is warm, have students add enough water to keep the material moist. After two weeks, ask them to check the compost. When the layered material has combined with the soil, invite students to use it in the experiment on pages 49 and 50.

Note: Do not add any meat or dairy products to the compost. The decomposition of these items causes a terrible odor and can attract vermin. Have students wash hands thoroughly after handling compost containers.

The Nitrogen Cycle

Soil Types

(Demonstration/Observation)

The earth's soil varies from region to region. Soil is a combination of mineral particles mixed with air, water, and living and dead organic matter. The mineral content in a region's soil helps determine its properties and its ability to sustain plant life. Mineral particles range in size from small to large—clay particles are small, silt particles are medium-size, and sand particles are large. To determine the type of soil found in different regions, have students conduct a jar test to identify mineral particles.

1. Collect samples of three different dry, crushed soils in three-quart jars. Put a one-inch layer of one soil into each jar. Label the jars: *A, B,* and *C.*
2. Fill the jars 2/3 full of water and add one teaspoon of Calgon® or table salt to each jar to act as a dispersing agent.
3. Close the jars and shake thoroughly.

Sand will settle to the bottom in one minute. Ask students to hold a ruler against the outside of the jars to measure the depth of each sandy layer. Silt will settle on top of the sand in about four to five hours. Have students measure the silt layers in the same way they measured the sandy layers. Clay may take a day or more to settle on top. If the water is cloudy, the clay requires more time to settle. When it settles, have students measure the clay layer. Once all layers have settled, ask students to explain which sample is mostly clay, sand, or silt; or which is a combination of the three. After discussion, have students use a map to hypothesize where each soil can be found naturally.

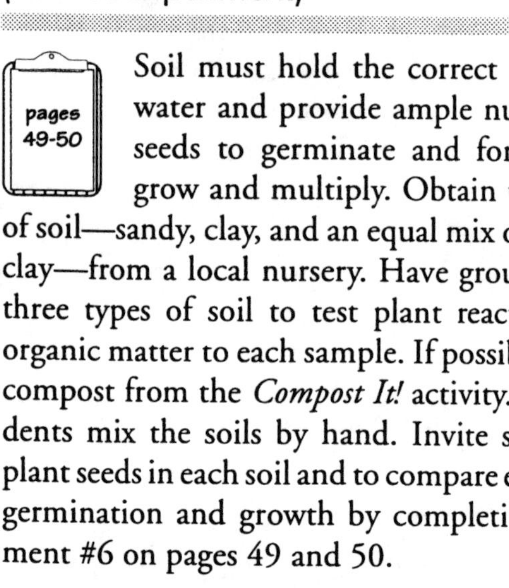

Soil for Growth

(Science Experiment)

Soil must hold the correct amount of water and provide ample nutrients for seeds to germinate and for plants to grow and multiply. Obtain three types of soil—sandy, clay, and an equal mix of sand and clay—from a local nursery. Have groups use the three types of soil to test plant reactions. Add organic matter to each sample. If possible, use the compost from the *Compost It!* activity. Have students mix the soils by hand. Invite students to plant seeds in each soil and to compare each plant's germination and growth by completing Experiment #6 on pages 49 and 50.

Name:

Soil for Growth
Experiment 6

EXPERIMENT 6

Question: ***Which soil type—sandy, clay-like, or a combination—is best for supporting plant life?***

Hypothesis: ______________________________

____

Materials

- ❑ equal amounts of three soil samples (sandy, mostly clay, and equal parts of sand and clay)
- ❑ three containers for planting
- ❑ three different types of seeds such as a legume, a gourd, a monocot, or a flowering plant (six of each type)
- ❑ water
- ❑ labels for containers
- ❑ tray to catch water runoff
- ❑ craft sticks

Procedure:

Step 1

Label the planting containers: *Sample A, Sample B,* and *Sample C.*

A. Mostly Sand
B. Mostly Clay
C. Sand/Clay

Step 2

Add half of one soil sample to each container, making sure the soil levels are equal.

Step 3

Place two seeds of each type in each container. Mark the name and location of the seeds by placing a craft stick with the seed name written on it in the soil next to the seeds. Add the rest of the soil to cover the seeds.

Step 4

Water every other day unless the environment is hot and dry, requiring daily watering. Water each container equally.

Step 5

Observe the containers each day for two to three weeks. Complete a table similar to the one shown.

Date	Amount of Water	Sprouted Seeds (Yes, No)	Height of Tallest Seedling

Name:

Soil for Growth
Experiment 6

EXPERIMENT 6

Results and Conclusions:

1. Describe any observations regarding the plants in soil samples *A, B,* and *C.* Why do you believe the plants grew as they did? ________

2. Look at your table. What did you observe about the growth of your seeds over time? ________

   What is your explanation? ________

3. Did all three types of seeds grow in the same way in each soil sample? Why or why not? ________

4. Which type of soil dried out most quickly? How do you explain this?

5. Which soil sample would be best for planting a garden? Why?

Science Challenge: Set up an experiment to test this question:

Can sandy soil be enriched to promote plant growth?

Write your question, hypothesis, procedure, and materials list on another sheet of paper. Then test the hypothesis and record your conclusions.

Soil and Land Exploitation

Life is dependent on the richness of the earth's soil and the abundance of its resources, yet people continue to **exploit** the land. Natural plant life is removed to make room for farming. Fertilizers and pesticides add harmful chemicals to the soil. Land is covered with concrete and asphalt. Mining strips away **topsoil** and dumps heavy metals into the soil. Mountains of trash load landfills to capacity. To change the negative effects of this abuse, people must gain a better understanding of land exploitation.

* *

Land Erosion

(Science Experiment)

Erosion is the movement of the earth's soil and rock due to weathering. Weathering, the breakdown and loosening of rock and soil, can be caused by ice, chemicals, wind, water, or heat from the sun. A lack of abundant plant life encourages weathering and erosion. Plant roots help hold soil in place. Roots absorb water and protect soil against the direct force of wind and rain. When plant life is removed to make room for farmland, construction, or mining, soil is exposed to erosion by wind, water, or other forces. Have students observe erosion in plantless soil by completing Experiment #7 on pages 52 and 53. After the experiment, invite students to look for examples of erosion in places such as vacant lots, hillsides, or the beach.

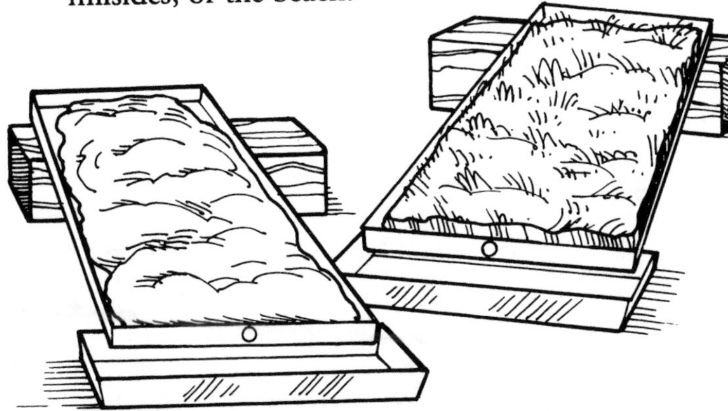

Strip Mining

(Demonstration)

Mining is the removal of sand, oil, natural gas, metal, coal, and other minerals from the earth. One form of mining, strip mining, is used to excavate minerals. Minerals are exposed when vast areas of soil are removed with heavy machinery. The excess soil is placed in piles where it is rendered unsuitable for plant growth. Demonstrate this process with the following activity:

1. Place a thin layer of blue clay in the bottom of a tray. Place some B-B's in the clay to represent a mineral.
2. Mold a 1" layer of yellow clay over the blue clay to represent subsoil.
3. Put a 1/4" layer of green clay over the yellow clay. The green layer represents the topsoil which contains nutrients for plant life.
4. Have students use paper clips as excavating tools to "mine" for minerals, digging out trenches. Pile up excess "soil" on the remaining surface of the model.

Since the blue layer is at the bottom, large amounts of "soil" will be piled up before students find the minerals. Once the minerals have been excavated, discuss what has happened to the topsoil, subsoil, and mineral-filled layer. Have students explain the consequences of strip mining for a natural environment and its plants and animals.

Name:

Land Erosion
Experiment 7

EXPERIMENT 7

Question: *Do plants help prevent erosion of the soil?*

Hypothesis: ______________________________________

____

Materials

- ❑ shallow aluminum pan filled with dirt
- ❑ shallow aluminum pan filled with sod
- ❑ two trays to collect runoff
- ❑ two blocks of wood at least two-inches thick
- ❑ two quart-size plastic containers with holes punched in the bottom
- ❑ two pitchers of water

Procedure:

Step 1

Place one end of each aluminum pan on a block of wood to form a slope.

Step 2

Poke a hole at the bottom end of the sloping trays to allow water to run off. Place a tray under each hole to catch the escaping water.

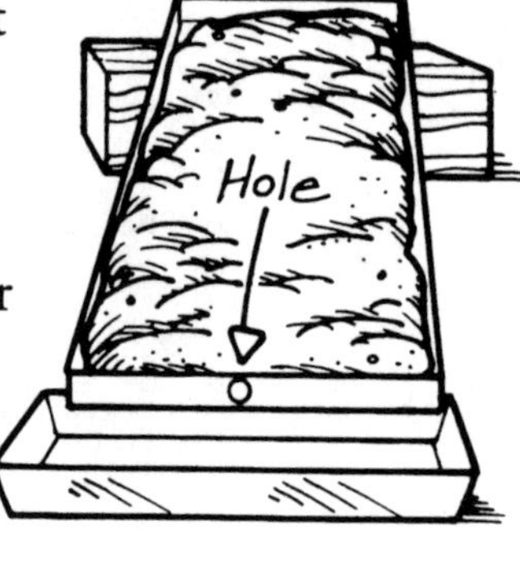

Step 3

Hold an empty quart container over the upper end of each slope. Quickly pour water into each quart container. Observe the pans as water "rains" out of the quart containers. Allow the same amount of water to flow down each tray.

Step 4

Observe both the dirt and sod pans and the trays of water at the bottom.

Name:

Land Erosion
Experiment 7

EXPERIMENT 7

Results and Conclusions:

1. Compare the sod pan to the dirt pan. What are the similarities and differences? ______________________________

2. What factors would affect the amount of erosion that could occur on each surface? ______________________________

3. Compare the water in the sod run-off tray to the dirt run-off tray. Are there any differences in the appearance of the water in each? Why or why not? ______________________________

4. Observe the appearance of the water. Over time, what would be the effects on the ability of the land to grow crops? Why? ______________________________

5. After a grass or forest fire, it is important to reseed the land before heavy rains. Based on your observation, why would this be necessary? ______________________________

Science Challenge: Set up an experiment to test this question:

Could the horizontal or vertical direction of crop rows affect the erosion on a hill?

Write your question, hypothesis, procedure, and materials list on another sheet of paper. Then test the hypothesis and record your conclusions.

Soil and Land Conservation

People are finding ways to maintain, preserve, and replenish the earth's soil and its resources. People can prevent the erosion of nutrient-rich topsoil by leaving habitats in their natural state, by planting vegetation between crop rows, and by making **windbreaks.** Nutrients can be replaced into the soil by planting special crops or by rotating crops. People are also learning to recycle and use proper composting techniques. In addition, scientists are searching for new energy sources which will reduce the need for burning fossil fuels.

* *

Lunchtime Waste Management

(Survey/Graphing)

To help students understand the amount and type of waste that is sent to the local landfill, have them monitor the trash accumulated during lunch hour. Have students spend a week observing and recording the types of waste left over from school lunches. Ask students to classify trash into *recyclables, compostables, reusables,* and *landfill items*. Have students spend one day collecting lunchtime trash. Ask students to physically group the items into each of the four categories. Have students brainstorm how the trash can be reduced and recycled. As a follow-up, challenge groups to implement a schoolwide campaign to reduce waste and recycle paper, plastic, and glass. After one month, have students evaluate the success of their program and make adjustments to insure greater success.

Note: When working with trash, be sure students wear gloves and handle metal and glass items properly.

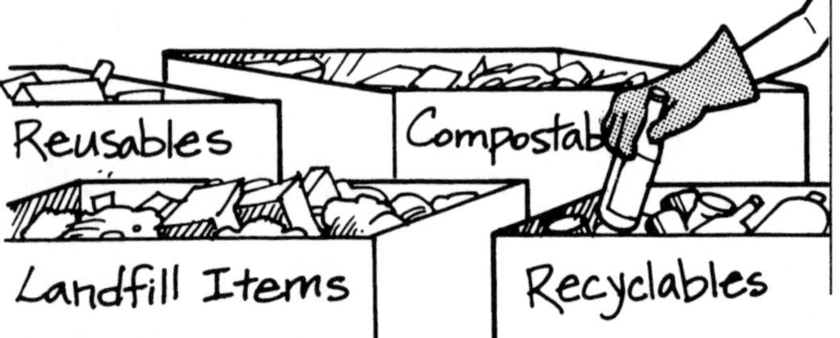

Land Use and Temperature

(Science Experiment)

A growing population must make intelligent choices about how limited land space is used. As we build cities and connect them with roads, we replace natural vegetation with buildings, asphalt, and concrete. This minimizes the amount of land available for food production. Concentrated areas of asphalt and concrete raise the air temperature which may contribute to global warming. If global warming becomes severe, ecosystems will be at risk. In Experiment #8 on pages 55 and 56, students observe temperature differences relative to the presence of soil, trees, asphalt, and concrete.

Name:

Land Use and Temperature
Experiment 8

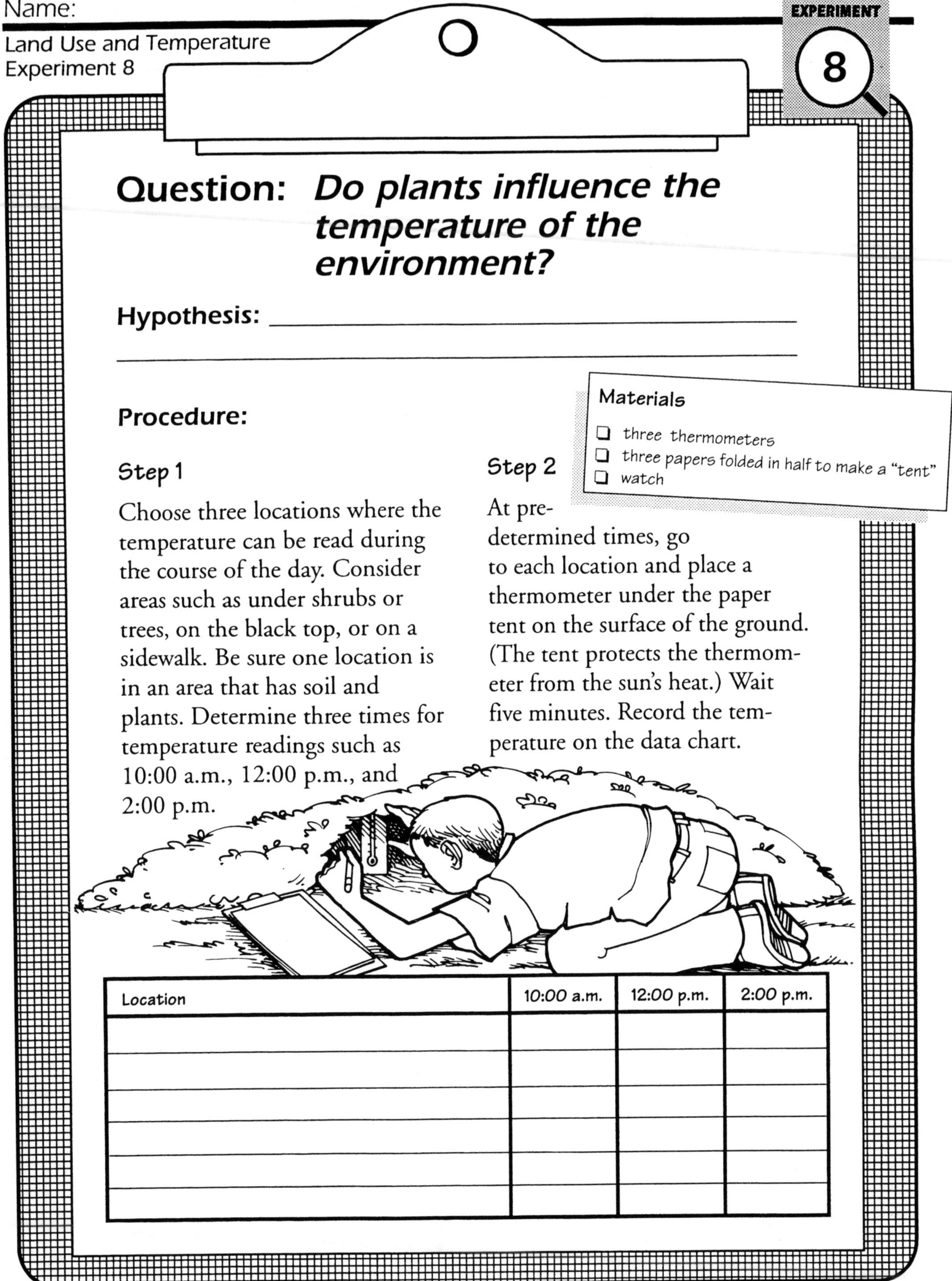

Question: *Do plants influence the temperature of the environment?*

Hypothesis: ______________________________

____

Procedure:

Materials

- ❑ three thermometers
- ❑ three papers folded in half to make a "tent"
- ❑ watch

Step 1

Choose three locations where the temperature can be read during the course of the day. Consider areas such as under shrubs or trees, on the black top, or on a sidewalk. Be sure one location is in an area that has soil and plants. Determine three times for temperature readings such as 10:00 a.m., 12:00 p.m., and 2:00 p.m.

Step 2

At predetermined times, go to each location and place a thermometer under the paper tent on the surface of the ground. (The tent protects the thermometer from the sun's heat.) Wait five minutes. Record the temperature on the data chart.

Location	10:00 a.m.	12:00 p.m.	2:00 p.m.

EXPERIMENT 8

Results and Conclusions:

1. Which location had the greatest change in temperature? ________

   ____

   Why do you think this was so? ________________________

   ____

   ____

2. Which location remained coolest throughout the day? ________

   ____

   Why do you think this was so? ________________________

   ____

   ____

3. Compare your group's results to those of your classmates. Were the results the same or different? How do you account for this? _____

   ____

   ____

   ____

4. Based on the results of this experiment, how can the growth of the human population influence the temperature of our planet?

   ____

   ____

   ____

5. Using the results from this experiment and what you know about the greenhouse effect, what dangers can you see for the earth?

   ____

   ____

   ____

Science Challenge: Set up an experiment to test this question:

Can plants help lower our city's temperature?

Write your question, hypothesis, procedure, and materials list on another sheet of paper. Then test the hypothesis and record your conclusions.

Extension Activities

Where the Forest Meets the Sea

(Art/Literature)

In the book *Where the Forest Meets the Sea*, a young boy and his grandfather go to a secluded location. The grandfather describes the place the way it was millions of years ago and speculates whether it will remain in its natural state in the future.

The author uses interesting textures and materials to create collages with hidden images that tell stories. Invite students to collect and bring in recycled, textured materials to create environmental scenes.

The Dust Bowl

(Research/Writing)

The Great Plains were once covered by native grasses that farmers plowed away to plant wheat and to create grazing land. In the 1930s, severe drought and heavy winds killed the farmers' crops, leaving the soil exposed. Consequently, 50 million acres of topsoil that contained nutrients necessary for plant growth were blown hundreds of miles away. Have students study this period for the effects this disaster had, not only on the land, but on the lives of the people involved. Ask students to assume the role of a family member whose life was affected by the catastrophe of the Dust Bowl and write a diary entry about the experience.

Paper Making

(Art/Science)

Invite students to bring in "junk mail" from home for two weeks. Have students weigh the mail and estimate how many pounds are sent to the entire school or town population. Show students how to recycle this trash by making paper. Use the recycled paper for notes, scrap paper, or invitations to the "Ecology Fair" described on page 58.

To make paper:

1. Provide student groups with a one-foot square of screen, an old blender, and junk mail (newsprint works best).
2. Have students tear the junk paper into small pieces, no larger than a dime, and soak the pieces in water overnight.
3. The next day, set up the blender and screens in an outside area. Have students place the paper and water in a blender and touch the pulse button until the paper is mush. Students can add bits of construction paper to intensify the color.
4. Have students pour and spread the mixture onto the screens. Invite students to press the paper against the screens until most of the water is squeezed out.
5. Have students turn the screens upside down to loosen the paper.
6. Leave the screens on newspapers to dry.
7. When the paper on the screens is dry, ask students to carefully remove it.

Culminating Activities

Ecology Fair

Plan a culmination day for the ecology unit. Have students brainstorm creative activities for the day such as:

- Invite parents or other classes to hear presentations about ecology concepts.
- Encourage students to collect and bring in interesting scraps of paper, metal, or plastic materials from the trash. Invite students to use the trash to create a game, tool, structure, art form, shelter, clothing item, or toy. Display the creations for visitors.
- Set up classroom booths where visitors can observe the results of experiments. Demonstrate items such as the aquifer or the water filter.
- Invite visitors to take a tour of the school to pick up any trash found along the way.
- Have students take visitors on a nature walk around the school, pointing out various ecological communities present in and around the school.

Once students have finalized a list of activities for the Ecology Fair, help them identify tasks and form teams. Provide time each day for groups to work on their projects. Use invitations and posters to advertise the special event. And, don't forget the video camera!

Guest Speakers

Due to greater ecological concern, there is a growing demand for environmentally conscious professions. Some professions, such as botanist, forester, or biologist, are in the field of science. Other careers include environmental protection specialist, toxicologist, and hazardous waste manager. Have students write to professionals in the field of ecology and invite them to speak to the class on one day. Invite other classes to listen to the speakers and participate in discussions. After the presentations, have students write letters about the professions they investigated, explaining why each is important and how it benefits the environment.

Interviews

Divide the class into groups. Ask each group to do research about the profession of one invited guest speaker. Have students use this research to formulate interview questions. After the guests have completed their presentations, have each group meet with the speaker who has the profession they studied. Have students use their questions to interview the speakers. When the interviews are complete, invite each group to report their findings to the class.

Wildlife Rehabilitation

Visit a wildlife rehabilitation center in your area. For information, contact your local office of the U.S. Department of Fish and Game listed in the phone book under U.S. Government. *All Wild Creatures Welcome* by Patricia Curtis lists wildlife rehabilitation centers by region within the United States.

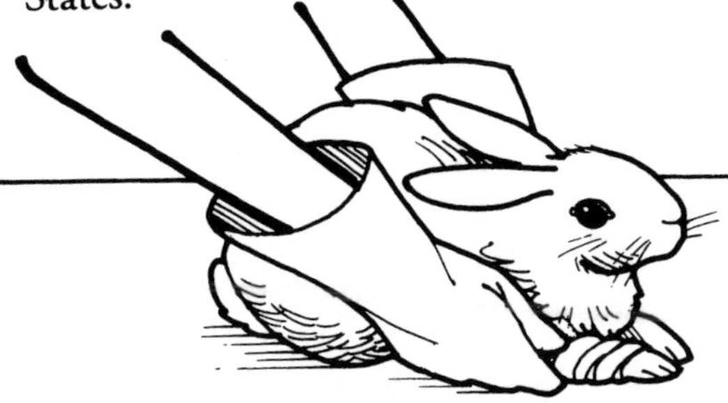

Career Interview

Name: ______________________

Interviewee: ______________________

____

Career Role: ______________________

1. How long have you been in this career? ______________________

2. What are your main responsibilities?
____
____
____

3. Why is your job important?
____
____
____

4. What kind of training or education was necessary before you were able to begin your career? ______________________
____

5. What is your most favorite and least favorite aspect of your job?
____
____
____
____

6. What are some interesting and/or important things you have learned during your time in this career?
____
____
____
____

Assessment

Ideally, science assessment includes evaluation of conceptual understanding, hands-on experiences, application of knowledge, and communication of learning. The following tools may assist you:

Science Portfolios—This is a collection of representive student work which may include science journals, scientific method and processes sheets (pages 6-7), and photos and videos of special student projects. You may also wish to include anecdotal records, summaries of parent/teacher communication, and performance evaluations (page 61).

Anecdotal Records—Keep written records of observations which verify students' understanding of the scientific method and processes during hands-on activities. Use these to assist in student and parent conferencing.

Student Conferences—Ask students to discuss their most interesting experiment. Guide them using questions like—

- Why did your experiment have those results?
- Did the results of your experiment support your hypothesis? Why or why not?
- What variables could have affected your results?
- What conclusions can you draw from the results?

Record student responses in writing or with a tape recorder.

Parent Conferencing—Discuss portfolios, student conferences, anecdotal records, and performance evaluations.

Performance Evaluations—Many teachers use experiment rubrics to evaluate individual student performance. Rubrics are paragraphs which describe benchmark levels of performance. The general Performance Evaluation on page 61 will help you give a numerical value for specific objectives addressed in many science rubrics. To obtain a rubric value, simply add the numbers and divide by 16. Or, better yet, create your own benchmark descriptions for the specific projects required in your classroom.

Experiment Rubric

Name: ______________________

Hypothesis: ______________________

Description	Value
Exemplary performance on both the experiment and written evaluation. Detailed and careful attention given tothe scientific method and data gathering. The conclusion demonstrates appropriate justification, insightful reasoning, and excellent critical thinking skills.	5
Experiment and written requirements completed correctly. Attention given to the steps of the scientific method. Able to support the conclusion with appropriate data.	4
Experiment complete but conclusion is not justified based upon data collected.	3
Experiment completed incorrectly.	2
Experiment incomplete.	1
No attempt made.	0

Name:

Performance Evaluation

Directions:
Circle the number which best reflects the frequency with which each behavior is observed.

0-never 1-rarely 2-occasionally 3-sometimes 4-often 5-frequently 6-always

Demonstrates understanding of important scientific concepts	0	1	2	3	4	5	6
Demonstrates understanding of the scientific method	0	1	2	3	4	5	6
Keeps accurate records of observations	0	1	2	3	4	5	6
Organizes data through categorizing and ordering	0	1	2	3	4	5	6
Draws logical conclusions from experiment results	0	1	2	3	4	5	6
Clearly communicates learning	0	1	2	3	4	5	6
Compares results of experiments with other similar experiments	0	1	2	3	4	5	6
Relates prior knowledge to new learning	0	1	2	3	4	5	6
Makes connections across the curriculum	0	1	2	3	4	5	6
Makes inferences	0	1	2	3	4	5	6
Applies knowledge to solve problems	0	1	2	3	4	5	6
Demonstrates appropriate use of lab equipment	0	1	2	3	4	5	6
Demonstrates curiosity	0	1	2	3	4	5	6
Is able to work cooperatively with others	0	1	2	3	4	5	6
All work is completed	0	1	2	3	4	5	6
Work is completed on time	0	1	2	3	4	5	6

Teacher Comments:

Student Comments:

Parent Comments:

Bibliography

Nonfiction

Baines, John. ***Acid Rain.*** Steck-Vaughn, 1989. Describes formation of acid rain, its effects on the environment, and efforts to control it.

Basta, Nicholas. ***The Environmental Career Guide.*** J. Wiley & Sons, 1991. Job opportunities with the earth in mind.

Becklake, Sue. ***Waste Disposal and Recycling.*** Gloucester, 1991. A look at the waste disposal industry, with an emphasis on recycling.

Bellamy, David. ***The River.*** Crown, 1988. The chronicle of an environmental accident on a river community and its impact on the animal and plant life.

Caduto, Michael J. and Joseph Bruchac. ***Keepers of the Earth.*** Fulcrum, 1989. Environmental issues and concerns explored through the telling of Native American stories.

Carr, Terry. ***Spill! The Story of the Exxon Valdez.*** Watts, 1991. Pictorial account of the 1989 oil tanker spill and subsequent clean-up efforts.

Cole, Joanna. ***Plants in Winter.*** HarperCollins, 1973. A description of the physiology of trees and their special adaptations developed to survive the winter.

Curtis, Patricia. ***All Wild Creatures Welcome.*** Dutton, 1985. The story of a wildlife rehabilitation center.

Goodman, Billy. ***A Kid's Guide to How to Save the Planet.*** Avon/Camelot, 1990. Discussion of environmental problems in the areas of waste, food, living things, and recycling.

Guiberson, Brenda Z. ***Cactus Hotel.*** Holt, 1991. Follows the saguaro cactus from its beginning as a tiny seed to its development as a shelter for many desert dwellers.

Hawkes, Nigel. ***Toxic Waste and Recycling.*** Gloucester, 1991. A look at toxic waste sites around the world, the difficulties in handling waste, and what is being done to clean up waste.

Herberman, Ethan. ***The City Kid's Field Guide.*** Simon & Schuster, 1989. An observation of an empty field, leading to many discoveries.

Huff, Barbara A. ***Greening the City Streets: The Story of Community Gardens.*** Clarion Books, 1990. Follows the efforts of a city to turn empty lots into gardens for its citizens to enjoy.

Hughey, Pat. ***Scavengers and Decomposers: The Clean-Up Crew.*** Macmillan, 1984. Descriptions of nature's efficient partners who break down organic waste into nutrients.

Miles, Betty. ***Save the Earth.*** Knopf, 1991. Amazing facts, kid's current projects, checklists, and a how-to section on ways kids can help the environment.

Mitchell, A. ***The Young Naturalist.*** Educational Development Corporation, 1984. Gives examples of how children can become young naturalists who take care of their environment.

Mutel, Cornelia and Mary M. Rodgers. ***Our Endangered Planet: Tropical Rain Forests.*** Lerner, 1991. Exploration of activities that endanger our rain forests and suggestions for conservation efforts.

Norsgaard, E. Jaediker. ***Nature's Great Balancing Act: In Our Own Backyard.*** Dutton, 1990. A general ecology book using the backyard environment as a reference.

Pedersen, Anne. ***The Kid's Environment Book: What's Awry and Why.*** John Muir Publications, 1991. A motivational book to make children aware of their environment and ways they can take care of the earth.

Pfeffer, Wendy. ***Popcorn Park Zoo: A Haven with a Heart.*** Julian Messner, 1992. The rescue and rehabilitation of wildlife at a zoo for abandoned animals.

Pollock, Steve. ***Ecology* (Eyewitness Books).** Knopf, 1993. A photographic look at the basics of ecology.

Pringle, Laurence. ***Chains, Webs, and Pyramids: The Flow of Energy in Nature.*** HarperCollins, 1975. Explanations of the ways in which energy is transferred through various food chains.

Pringle, Laurence. ***Living Treasure: Saving Earth's Threatened Biodiversity.*** Morrow, 1991. Exploration of endangered and extinct life and ways people can protect species currently threatened.

Reef, Catherine. ***Jacques Cousteau: Champion of the Sea.*** Holt, 1992. Examines the life of the French oceanographer.

Reef, Catherine. ***Rachel Carson: The Wonder of Nature.*** Holt, 1992. A look at the life of the admired biologist and conservationist.

Ecology Resources

Sinclaie, Patti (ed.). ***E for Environment: An Annotated Bibliography of Children's Books with Environmental Themes.*** Bowker, 1992. An excellent resource for classroom teachers.

Stanley, Jerry. ***Children of the Dust Bowl.*** Crown, 1992. The true story of the children of Weed Patch Camp in the Dust Bowl in the 1930s.

Talmadge, Katherine S. ***John Muir: At Home in the Wild.*** Holt, 1993. Follows the life of John Muir.

Fiction

Baker, Jeanne. ***Where the Forest Meets the Sea.*** Greenwillow Books, 1987. A young boy and his father make a trip to the Australian rain forest.

Cherry, Lynne. ***The Great Kapok Tree.*** Harcourt Brace, 1990. The animals of the rain forest convince a man with an ax to spare their home.

Cherry, Lynne. ***A River Ran Wild.*** Harcourt Brace, 1992. The environmental history of the Nashua River.

Clements, Andrew. ***Big Al.*** Picture Book Studio, 1988. Picture book about a lovable but strange-looking animal.

Freeman, Don. ***The Seal and the Slick.*** Viking, 1974. Fictional tale of the life of a seal pup saved by children.

Jeffers, Susan. ***Brother Eagle, Sister Sky: A Message from Chief Seattle.*** Dial, 1991. The words of Chief Seattle, adapted by the author, to share his message of respect for the earth.

Peet, Bill. ***The Wump World.*** Houghton Mifflin, 1970. Newcomers invade and pollute Wump World.

Seuss, Dr. ***The Lorax.*** Random House, 1971. The imaginary Once-ler describes the results of a local pollution problem.

Wood, Nancy. ***Spirit Walker.*** Doubleday, 1993. A collection of poems which express the values of Native Americans and their interconnectedness to the earth.

Technology

Video

Save the Earth: A How-to Video

International Video Publications in association with:
Save the Earth Brigade
Tri-Coast International
1020 Pico Blvd.
Santa Monica, CA 90405
60 mins. $19.95.

This inexpensive video offers ideas to reuse, reduce, and recycle.

Software

Acid Rain

Too Much Trash!

What's in Our Water?

National Geographic
MAC, Windows, DOS

Three software programs that are a part of the National Geographic Society's Kids Network program. This telecommunications software allows students to explore ecology topics with classes from around the nation.

Hardware requirements include a modem and phone line.

Free Materials and Catalogues

Smithsonian Resource Guide for Teachers

A catalogue of information on ordering free or inexpensive books, posters, audio tapes, videotapes, and other items for your classroom is available from The Smithsonian by writing to:

Elementary and Secondary Education
Arts and Industries Bldg.
Room 1163
MRC 402
Smithsonian Institution
Washington, D.C. 20560
(202) 357-2425

Glossary of Ecology Terms

acid rain—rain that picks up chemicals from polluted air and becomes acidic
adaptations—changes in organisms to make them better able to survive in their environment
air quality—the degree of purity concerning air conditions
algae—simple plants that live in water
aquifer—underground water reservoirs
atmosphere—the invisible blanket of air that surrounds the earth
biological accumulation—the process by which poisons accumulate in higher order organisms through the consumption of lower order organisms
biome—a major regional community such as the tundra, grasslands, or desert
biosphere—life-sustaining regions of the earth
carbon dioxide—a colorless, odorless gas made of carbon and oxygen
community—all plants and animals living in the same environment
compost—organic material that decomposes to return nutrients to the soil
compound—an element combined with other elements
condense (condensation)—to change water from a gas to a liquid through cooling
conserve (conservation)—to maintain, restore, or preserve
consumer—an organism that eats a living thing for energy
decompose—to break down into basic elements
decomposer—an organism that gets energy from decomposing material
ecosystem—communities interacting in their environment and functioning as a system
element—a basic substance which is a component of matter
emulsification—suspension of small molecules of one liquid on another liquid
environment—an organism's total surroundings
erosion—movement of earth and rock due to weathering
evaporate (evaporation)—to change water from a liquid to a gas through heating
filter (filtration)—to send a liquid through a porous surface to purify it
food chain—the order in which food energy is transferred through organisms, from plant to herbivore to carnivore
food web—a system of interconnecting food chains
fossil fuel—a substance originating from decayed organic material that is burned for energy
global warming—a theory that states the earth's atmosphere is warming due to trapped heat energy from the sun
ground water—water stored underground in soil
habitat—an environment where an organism lives
landfill—a huge pit for burying trash and garbage
nitrogen—a basic element found in the air, soil, and all proteins
nitrogen cycle—the transfer of nitrogen from the atmosphere into the soil and along the food chain
nutrient—something that nourishes, repairs, or promotes growth
organic—derived from a living organism
organism—a living thing
oxygen—the colorless, odorless, tasteless gas produced by plants
ozone layer —a form of oxygen (O_3) that surrounds the earth and protects it from ultra-violet rays
pesticide—pest-killer made from synthetic or organic material
phosphate—a chemical containing phosphoric acid
photosynthesis—the process by which plants use sunlight, water, and carbon dioxide to produce glucose and oxygen
plankton—drifting microscopic organisms found in water
pollute (pollution)—to make dirty and not healthy, to contaminate
population—all living things of the same species living in one area
precipitation—condensed water vapor in the form of rain, sleet, snow, and/or hail
predator—an animal who feeds on other animals
prey—an animal hunted as food
producer—an organism that makes its own food
range of tolerance—the minimum and maximum requirements an organism needs to live
resource—a natural source of supply for living things in an ecosystem
respiration—the process of absorbing and utilizing oxygen and releasing carbon dioxide
run-off channels—reservoirs for excess water to protect land from flooding
silt—sedimentary soil with a consistency between sand and clay
solvent—a material that can dissolve another substance
strip mining— a technique by which vast areas of soil are removed to extract minerals
thermal pollution—the release of waste water at higher-than-normal temperatures
topsoil—surface soil containing most of the nutrients required for plant growth
water cycle—the circulation of the earth's water through evaporation, condensation, precipitation, and accumulation
water vapor—water as a gas in the atmosphere